AF361772

Progress in Arrhythmia: Insights in Integrated Care, Anticoagulation Therapy and Precision Medicine

Progress in Arrhythmia: Insights in Integrated Care, Anticoagulation Therapy and Precision Medicine

Editor

José Miguel Rivera-Caravaca

MDPI • Basel • Beijing • Wuhan • Barcelona • Belgrade • Manchester • Tokyo • Cluj • Tianjin

Editor
José Miguel Rivera-Caravaca
Faculty of Nursing,
University of Murcia,
Murcia, Spain

Editorial Office
MDPI
St. Alban-Anlage 66
4052 Basel, Switzerland

This is a reprint of articles from the Special Issue published online in the open access journal *Journal of Personalized Medicine* (ISSN 2075-4426) (available at: https://www.mdpi.com/journal/jpm/special_issues/arrhythmia_precision_medicine).

For citation purposes, cite each article independently as indicated on the article page online and as indicated below:

LastName, A.A.; LastName, B.B.; LastName, C.C. Article Title. *Journal Name* **Year**, *Volume Number*, Page Range.

ISBN 978-3-0365-8354-9 (Hbk)
ISBN 978-3-0365-8355-6 (PDF)

Contents

Editorial

Progressions in Cardiac Arrhythmia: Specific Populations and the Need for Precision Medicine

José Miguel Rivera-Caravaca [1,2,*] and Jeroen M. Hendriks [3,4]

1   Faculty of Nursing, University of Murcia, 30120 Murcia, Spain
2   Department of Cardiology, Hospital Clínico Universitario Virgen de la Arrixaca, Instituto Murciano de Investigación Biosanitaria (IMIB-Arrixaca), Centro de Investigación Biomédica en Red—Enfermedades Cardiovasculares (CIBERCV), 30120 Murcia, Spain
3   Caring Futures Institute, College of Nursing and Health Sciences, Flinders University, Sturt Road, Bedford Park South Australia 5042, GPO Box 2100, Adelaide, SA 5001, Australia; jeroen.hendriks@flinders.edu.au
4   Centre for Heart Rhythm Disorders, University of Adelaide and Royal Adelaide Hospital, Port Road, Adelaide, SA 5001, Australia
*   Correspondence: josemiguel.rivera@um.es; Tel./Fax: +34-868-88-18-06

Citation: Rivera-Caravaca, J.M.; Hendriks, J.M. Progressions in Cardiac Arrhythmia: Specific Populations and the Need for Precision Medicine. *J. Pers. Med.* **2023**, *13*, 1122. https://doi.org/10.3390/jpm13071122

Received: 5 July 2023
Accepted: 7 July 2023
Published: 11 July 2023

Atrial fibrillation (AF) is the most common cardiac arrhythmia in the general population. The high prevalence of this condition—which is expected to be even higher in upcoming years [1]—has motivated in-depth studies, from mechanistic and pathophysiological approaches to further understand how AF is initiated and perpetuated to analyses of the role and impact of integrated care in its management. In this Special Issue of the *Journal of Personalized Medicine*, some controversial and scarcely investigated topics related to AF have been introduced, with the aim of providing holistic care and treatment in AF. These topics include improving the detection, adverse outcome prevention, and risk factor optimization in AF patients.

Escribano et al. aimed to conduct an analysis of the f-wave harmonic spectral structure to improve catheter ablation (CA) outcome prediction through several entropy-based measures computed on different frequency bands in 151 patients with persistent AF under radio frequency CA. Remarkably, the authors demonstrated in this pioneering analysis of the f-wave harmonic spectral structure that the presence of larger harmonics and a proportionally smaller dominant frequency peak was strongly associated with a decreased probability of AF recurrence after CA [2].

To better understand the CA process, Vraka and colleagues have decomposed crucial CA steps, including coronary sinus catheterization and the impact of left and right pulmonary vein isolation (PVI). In brief, they showed that left PVI is the critical part of the CA of PVs for paroxysmal AF patients, significantly altering the P-wave duration, whereas the effect of the CA of PVs on the coronary sinus is less straightforward and is demonstrated to a lesser extent. The authors suggest that other atrial structures may be more indicative of the ablation outcome and should be assessed as alternative references [3].

Osorio et al. compared the performance of cycle length estimations of three different local activation wave detection methods: the hyperbolic tangent (HT) function, an adaptive amplitude threshold (AAT), and a cycle length iteration (ACLI). For the HT method, the accuracy, sensitivity, and precision were higher compared to the AAT and ACLI methods, with even a lower cycle length error. The authors concluded that the high robustness and precision demonstrated by the HT method promote its implementation on CA mapping devices for a more successful location of ablation targets and improving the results of CA procedures [4].

Currently, left atrial (LA) appendage surgical exclusion or percutaneous occlusion have limitations. For this reason, an interesting study by Pasta et al. sought to quantify the hemodynamic and structural loads of a novel potential procedure to partially invert the "dead" LA appendage space to eliminate the auricle apex, called LA appendage inversion

(LAAI). This procedure was simulated by pulling the elements at the LA appendage tip and prescribing a displacement motion along a predefined path. Later, the authors used the deformed configuration to develop a computational flow analysis of LAAI, demonstrating that the inverted LAA wall in the inverted appendage undergoes a change in the stress distribution from tensile to compressive, and this can lead to resorption of the LAA tissue as per a reduced stress/resorption relationship [5].

From a clinical point of view, three original studies performed in-depth analyses of the management of AF. Merino-Merino and collaborators, in a confirmatory study including patients with AF and healthy controls, showed that NT-proBNP, high sensitivity troponin T, and ST2, were significantly related to the presence of AF. Among these biomarkers of cardiac dysfunction, inflammation, and damage, NT-proBNP exhibited the best discrimination ability (with an area under the curve of 0.995) [6]. Pastori et al. identified clinical phenotypes of AF patients to stratify the mortality risk by a cluster analysis performed on 5171 AF patients from the nationwide START (survey on anticoagulated patients register). In brief, cluster 3, characterized by men with diabetes, coronary disease, and peripheral artery disease, and cluster 4, including mainly elderly women with previous cerebrovascular events, had a higher risk of mortality (aHR 6.702, 95% CI 2.433–18.461, and aHR 8.927, 95% CI 3.238–24.605; respectively) [7]. Roldán et al. presented the first study analyzing the impact of adherence to the Atrial Fibrillation Better Care (ABC) pathway on the quality of anticoagulation control in a cohort of AF outpatients starting vitamin K antagonists. Amongst ABC pathway non-adherent patients, a greater proportion had TTR < 65% or TTR < 70%, with a lower mean TTR in non-adherent patients ($59.4 \pm 22.3\%$ vs. $63.9 \pm 21.1\%$; $p = 0.004$). Indeed, adherence to the ABC pathway was independently associated with a TTR of $\geq$65% [8].

Additionally, two reviews aimed to shed light on very specific patient populations and therapeutic interventions. In the first review, Badescu and colleagues summarized the current evidence concerning the management of AF in people with hemophilia, which is a complex combination in terms of the prevention of bleeding and thromboembolic complications. These patients are often perceived as having a high bleeding risk due to coagulation factor VIII/IX deficiency, but in the presence of AF they also have a high risk of stroke and systemic embolism. Despite the lack of research in this population, it should be highlighted that the treatment offered to the general population, including CA and LA appendage occlusion, can be implemented in hemophiliacs if an appropriate replacement therapy can be provided [9]. Buckley et al. reviewed, synthesized, and proposed an AF-specific exercise rehabilitation guideline based on data from primary trials and real-world cohort studies. According to their analysis, a minimum exercise threshold of 360–720 metabolic equivalent minutes/week, corresponding to 60–120 min of exercise per week at moderate-to-vigorous intensity, could be an evidence-based recommendation for patients with AF. This would improve AF-specific outcomes, quality of life, and possibly prevent long-term major adverse cardiovascular events. Non-traditional, low–moderate intensity exercise, such as yoga, appears to have promising benefits to patient quality of life and should therefore be considered in a personalized rehabilitation program [10].

Finally, AF is generally underdiagnosed, mainly because it can be present without signs and symptoms. For this reason, the real prevalence of AF is assumed to be higher than the known prevalence. In this Special Issue, Jones et al. described the protocol of an interesting study with the objective of investigating the effectiveness of a hand-held device embedded into the handles of supermarket trolleys in screening for AF in the general population. This will be a mixed method two-phase study. The quantitative first phase will be a cross-sectional observational study involving participants from a convenience sample at four large supermarkets with pharmacies. The prospective participants will be asked to conduct their shopping using a trolley embedded with a MyDiagnostick single-lead electrocardiogram sensor. If the device identifies a participant with AF, the in-store pharmacist will be dispatched to take a manual pulse measurement and a static control sensor reading and offer a cardiologist consultation referral. In the qualitative second phase, semi-structured interviews carried out with those pharmacists and store managers during the running of the trial period will be performed to explore the perceptions of staff regarding the benefits of the device [11].

In summary, the studies included in this Special Issue cover different areas in the field of AF and provide an updated overview of the complexities of AF management. These novel insights are of great interest to healthcare professionals treating patients with AF and may assist in decision-making processes.

Conflicts of Interest: The authors declare no conflict of interest.

References

1. Kornej, J.; Börschel Christin, S.; Benjamin Emelia, J.; Schnabel Renate, B. Epidemiology of Atrial Fibrillation in the 21st Century: Novel Methods and New Insights. *Circ. Res.* **2020**, *127*, 4–20. [CrossRef] [PubMed]
2. Escribano, P.; Ródenas, J.; García, M.; Arias, M.A.; Hidalgo, V.M.; Calero, S.; Rieta, J.J.; Alcaraz, R. Preoperative Prediction of Catheter Ablation Outcome in Persistent Atrial Fibrillation Patients through Spectral Organization Analysis of the Surface Fibrillatory Waves. *J. Pers. Med.* **2022**, *12*, 1721. [CrossRef] [PubMed]
3. Vraka, A.; Bertomeu-González, V.; Fácila, L.; Moreno-Arribas, J.; Alcaraz, R.; Rieta, J.J. The Dissimilar Impact in Atrial Substrate Modificationof Left and Right Pulmonary Veins Isolation after Catheter Ablation of Paroxysmal Atrial Fibrillation. *J. Pers. Med.* **2022**, *12*, 462. [CrossRef] [PubMed]
4. Osorio, D.; Vraka, A.; Moreno-Arribas, J.; Bertomeu-González, V.; Alcaraz, R.; Rieta, J.J. Comparative Study of Methods for Cycle Length Estimation in Fractionated Electrograms of Atrial Fibrillation. *J. Pers. Med.* **2022**, *12*, 1712. [CrossRef] [PubMed]
5. Pasta, S.; Guccione, J.M.; Kassab, G.S. Inversion of Left Atrial Appendage Will Cause Compressive Stresses in the Tissue: Simulation Study of Potential Therapy. *J. Pers. Med.* **2022**, *12*, 883. [CrossRef] [PubMed]
6. Merino-Merino, A.; Saez-Maleta, R.; Salgado-Aranda, R.; AlKassam-Martinez, D.; Pascual-Tejerina, V.; Martin-Gonzalez, J.; Garcia-Fernandez, J.; Perez-Rivera, J.-A. A Differential Profile of Biomarkers between Patients with Atrial Fibrillation and Healthy Controls. *J. Pers. Med.* **2022**, *12*, 1406. [CrossRef] [PubMed]
7. Pastori, D.; Antonucci, E.; Milanese, A.; Menichelli, D.; Palareti, G.; Farcomeni, A.; Pignatelli, P.; the START2 Register Investigators. Clinical Phenotypes of Atrial Fibrillation and Mortality Risk—A Cluster Analysis from the Nationwide Italian START Registry. *J. Pers. Med.* **2022**, *12*, 785. [CrossRef] [PubMed]
8. Roldán, V.; Martínez-Montesinos, L.; López-Gálvez, R.; García-Tomás, L.; Lip, G.Y.H.; Rivera-Caravaca, J.M.; Marín, F. Relation of the 'Atrial Fibrillation Better Care (ABC) Pathway' to the Quality of Anticoagulation in Atrial Fibrillation Patients Taking Vitamin K Antagonists. *J. Pers. Med.* **2022**, *12*, 487. [CrossRef] [PubMed]
9. Badescu, M.C.; Badulescu, O.V.; Butnariu, L.I.; Floria, M.; Ciocoiu, M.; Costache, I.-I.; Popescu, D.; Bratoiu, I.; Buliga-Finis, O.N.; Rezus, C. Current Therapeutic Approach to Atrial Fibrillation in Patients with Congenital Hemophilia. *J. Pers. Med.* **2022**, *12*, 519. [CrossRef] [PubMed]
10. Buckley, B.J.R.; Risom, S.S.; Boidin, M.; Lip, G.Y.H.; Thijssen, D.H.J. Atrial Fibrillation Specific Exercise Rehabilitation: Are We There Yet? *J. Pers. Med.* **2022**, *12*, 610. [CrossRef] [PubMed]
11. Jones, I.D.; Lane, D.A.; Lotto, R.R.; Oxborough, D.; Neubeck, L.; Penson, P.E.; Czanner, G.; Shaw, A.; Smith, E.J.; Santos, A.; et al. Supermarket/Hypermarket Opportunistic Screening for Atrial Fibrillation (SHOPS-AF): A Mixed Methods Feasibility Study Protocol. *J. Pers. Med.* **2022**, *12*, 578. [CrossRef] [PubMed]

Journal of
Personalized Medicine

Article

Preoperative Prediction of Catheter Ablation Outcome in Persistent Atrial Fibrillation Patients through Spectral Organization Analysis of the Surface Fibrillatory Waves

Pilar Escribano [1,*], Juan Ródenas [1], Manuel García [1], Miguel A. Arias [2], Víctor M. Hidalgo [3], Sofía Calero [3], José J. Rieta [4] and Raúl Alcaraz [1]

[1] Research Group in Electronic, Biomedical and Telecommunication Engineering, University of Castilla-La Mancha, 02071 Albacete, Spain
[2] Cardiac Arrhythmia Department, Complejo Hospitalario Universitario de Toledo, 45007 Toledo, Spain
[3] Cardiac Arrhythmia Department, Complejo Hospitalario Universitario de Albacete, 02006 Albacete, Spain
[4] BioMIT.org, Electronic Engineering Department, Universitat Politecnica de Valencia, 46022 Valencia, Spain
[*] Correspondence: pilar.escribano@uclm.es

Abstract: Catheter ablation (CA) is a commonly used treatment for persistent atrial fibrillation (AF). Since its medium/long-term success rate remains limited, preoperative prediction of its outcome is gaining clinical interest to optimally select candidates for the procedure. Among predictors based on the surface electrocardiogram, the dominant frequency (DF) and harmonic exponential decay (γ) of the fibrillatory waves (f-waves) have reported promising but clinically insufficient results. Hence, the main goal of this work was to conduct a broader analysis of the f-wave harmonic spectral structure to improve CA outcome prediction through several entropy-based measures computed on different frequency bands. On a database of 151 persistent AF patients under radio-frequency CA and a follow-up of 9 months, the newly introduced parameters discriminated between patients who relapsed to AF and those who maintained SR at about 70%, which was statistically superior to the DF and approximately similar to γ. They also provided complementary information to γ through different combinations in multivariate models based on lineal discriminant analysis and report classification performance improvement of about 5%. These results suggest that the presence of larger harmonics and a proportionally smaller DF peak is associated with a decreased probability of AF recurrence after CA.

Keywords: persistent atrial fibrillation; catheter ablation; outcome prediction; fibrillatory wave analysis; electrocardiogram; spectral analysis; dominant frequency; harmonic content

Citation: Escribano, P.; Ródenas, J.; García, M.; Arias, M.A.; Hidalgo, V.M.; Calero, S.; Rieta, J.J.; Alcaraz, R. Preoperative Prediction of Catheter Ablation Outcome in Persistent Atrial Fibrillation Patients through Spectral Organization Analysis of the Surface Fibrillatory Waves. *J. Pers. Med.* **2022**, *12*, 1721. https://doi.org/10.3390/jpm12101721

Academic Editor: José Miguel Rivera-Caravaca

Received: 20 August 2022
Accepted: 11 October 2022
Published: 14 October 2022

1. Introduction

Atrial fibrillation (AF) is the most frequently encountered cardiac arrhythmia in clinical practice [1]. This disruption of heart sinus rhythm (SR) affects roughly 37.5 million people worldwide [2], thus making it one of the most important public health problems and a significant cause of rising healthcare costs in developed countries [3]. Although the mechanisms underlying AF are not fully understood, it requires a combination of triggers, mainly located near the pulmonary veins (PVs), and a vulnerable atrial substrate characterized by reentry circuits [4]. AF is commonly linked to life-altering symptoms such as palpitations, fatigue, chest pain, shortness of breath, and dizziness [5]. This arrhythmia is not a direct cause of death, but it is associated with a two-fold increase in mortality because of a higher risk of heart failure and ischemic stroke [6].

Depending on the duration and recurrence of arrhythmic episodes, AF is classified into several stages [7]. However, the disease is not static and often progresses from paroxysmal to sustained forms over the course of a few years; this is usually correlated with irreversible electrical and structural remodeling of the atrial substrate that, in turn, promotes perpetuation of the arrhythmia [4]. Hence, early diagnosis of AF and finding the best way to restore SR as soon as possible is highly advisable [8]. To that purpose, pharmacological or electrical cardioversion strategies are frequently used in clinical practice until they are no longer effective or not recommended. In that case, catheter ablation (CA) has become a commonly used alternative [9] thanks to its superior ability over antiarrhythmic drugs for restoring SR in the midterm [10]. This therapy has evolved during the last two decades, and the most common protocol remains based on PVs isolation (PVI) [11]. Despite its usually successful initial outcome, it is not as effective in patients with persistent AF compared to patients with paroxysmal AF in the medium/long term, since approximately 40% to 50% of them relapse to AF within the first year [9]. Thus, the high complexity in the management of catheters and the complications associated with the procedure encourage careful assessment of the benefits and risks for each patient [12].

This tailored assessment, along with the clinical, social, and economic challenges that AF will involve in the coming decades [2], motivate the emerging clinical interest in the development of preoperative predictors of CA outcome, which could be helpful in the selection of patients who would benefit the most from the procedure [13]. Some benefits of choosing optimal candidates for CA are the reduction of hospitalization rates, limitation of repeated procedures, assignment of more appropriate AF treatments to improve patient quality of life, and minimization of the risks and costs associated with AF treatment [13]. So far, some clinical indices have been proposed to anticipate midterm CA outcome, such as the total duration of AF, the duration of the last arrhythmic episode, left atrial diameter, and history of hypertension or diabetes, among others. However, most of these parameters have reported controversial results, so there is no strong evidence to consider them as reliable single predictors of AF recurrence after the procedure [14].

Some authors have pioneered the use of parameters manually measured on the surface electrocardiogram (ECG) obtained prior to CA. Uncoordinated electrical conduction in the atria means the typical P-wave is replaced on the ECG by undulatory activity known as fibrillatory waves (f-waves). Manual measurement of the time between these waves, i.e., the AF cycle length (AFCL) [15], and their amplitude [16,17] have been positively correlated with midterm CA results. During ablation, invasive measures of the AFCL or its inverse, i.e., the dominant frequency (DF), in different atrial structures have also been confirmed to have a direct link with AF recurrence several months after the procedure [18–21]. However, these indices have the disadvantage of being based on manual and/or invasive ECG metrics, which entails a high degree of subjectivity in measurement during or after CA intervention.

To overcome these limitations and to achieve predictors available for the selection of candidates for CA, more recent studies have addressed non-invasive extraction of the aforementioned metrics through automated processing of the preoperative surface ECG. Researchers have reported performance at least as good as previously proposed clinical parameters and their manually or invasively derived versions [22]. Beyond the single DF and f-wave amplitude (FWA), the presence of large harmonics of the DF has also been identified as a promising predictor of CA outcome [18,23]. Higher harmonic content in the f-wave spectral distribution has been associated with a greater degree of organization of the atrial electrical activity and with higher probability of maintaining SR after different AF treatments, including CA [18,23], electrical cardioversion [24,25], and pharmacological therapy [26]. However, only the power contained by the DF and its harmonics has been analyzed to date, and the main goal of the present work is, hence, to conduct a broader analysis of the f-wave harmonic spectral structure to improve preoperative prediction of CA outcome in persistent AF patients.

2. Materials and Methods

2.1. Study Population

The population enrolled in this work consisted of 151 AF patients (35 women and 116 men) between 20 and 82 years old, with 59 years being the rounded mean age of the group. They were consecutively treated with radiofrequency CA for the first time following standard clinical indications at two Spanish hospitals (i.e., University Hospitals of Toledo and Albacete), thus constituting a retrospective database. The ablation procedure started with patient sedation using general anesthesia or conscious sedation after suspending all antiarrhythmic drug therapy except amiodarone >5 half-lives before the intervention. Anticoagulant drugs were also used to avoid thromboembolic complications. Indeed, an initial bolus of heparin was administered, followed by additional doses properly activated by coagulation-time monitoring throughout the procedure. Isolation of PVs was achieved by creating electrically impenetrable boundaries surrounding their ostia using a radiofrequency source [27]. Thus, a catheter was used to generate point-by-point lesions through the release of radiofrequency current for at least 30 s and then creating a contiguous antral circumferential line around the PVs, whose location were determined using a mapping catheter [28]. The procedure finished when all PVs were successfully isolated. If AF still remained at that point, SR was restored by electrical cardioversion.

The procedure was initially successful in all patients, who were monitored for several hours after the intervention without presenting any important complications. They received anticoagulant and antiarrhythmic drugs according to clinical judgment and had a quite standard follow-up with a visit, ECG, and 24 h Holter at 3 and 9 months, as recommended by the current 2020 ESC guidelines [7]. Thereafter, visits each 12 months were planned. A blanking period of 3 months was considered to define every episode of AF or other arrhythmias lasting more than 30 s as recurrence, but the patients were advised to go at any moment to an emergency room in case of AF-related symptoms.

2.2. Data Preprocessing

The database of this study consisted of a standard 12-lead ECG signal continuously recorded just before starting the CA procedure while the patient was under AF. Hence, a total of 151 ECG recordings with variable durations of between approximately 6 s and 5 min were available. The signals were acquired by the equipment available in the hospitals, with 16-bit resolution and a sampling rate of 977 Hz. For the present study, lead V1 was selected since it contains the f-waves with the greatest amplitude regarding ventricular activity, thus favoring their automated extraction and accurate analysis. In fact, the electrode that records this unipolar lead is located in a standard position very close to the atria, so it is the most appropriate lead for recording AF activity [29].

The selected ECG signal was then preprocessed to minimize the perturbations unavoidably recorded together with the electrical heart activity and, consequently, to improve further analysis of the f-waves. Since the patient was at rest during recording of the surface ECG before CA, the main disturbances presented on the ECG consisted of baseline wander, powerline interference, and disturbances associated with high-frequency noise sources [30]. The low-frequency component associated with baseline wander was estimated using a low-pass filter with a cut-off frequency of 0.8 Hz and zero-phase distortion through an IIR structure and forward/backward filtering and then subtracted from the original ECG signal [30]. Powerline interference was removed by a denoising algorithm based on the stationary wavelet transform. It applies a novel threshold function to the wavelet coefficients for filtering the unwanted frequency and its harmonics, simultaneously minimizing the distortion of the rest of the components [31]. The algorithm was also able to remove most of the high-frequency noise, but a low-pass forward/backward IIR filter with a cut-off frequency of 70 Hz was additionally used to make the ECG signal as clean as possible [30].

Finally, f-wave characterization requires that ventricular activity is first canceled. Although there are different techniques for this, f-waves were automatically extracted from the preprocessed ECG signal using a well-established QRST cancellation technique based on adaptive singular value cancellation (ASVC). The algorithm firstly obtained a cancellation template from the singular-value decomposition of a set of QRST complexes that were temporally aligned with regard to the R-peak. Then, the resulting template was adapted in amplitude for the cancellation of each single QRST complex. Notably, this method avoids spikes caused by discontinuities between QRST complexes and the subsequent TQ intervals through a softening approach that considers the differences between the cancellation template and the QRST segment at its beginning and end to minimize sudden transitions [32]. As an example, Figure 1 shows the *f*-waves obtained from a typical preprocessed ECG interval.

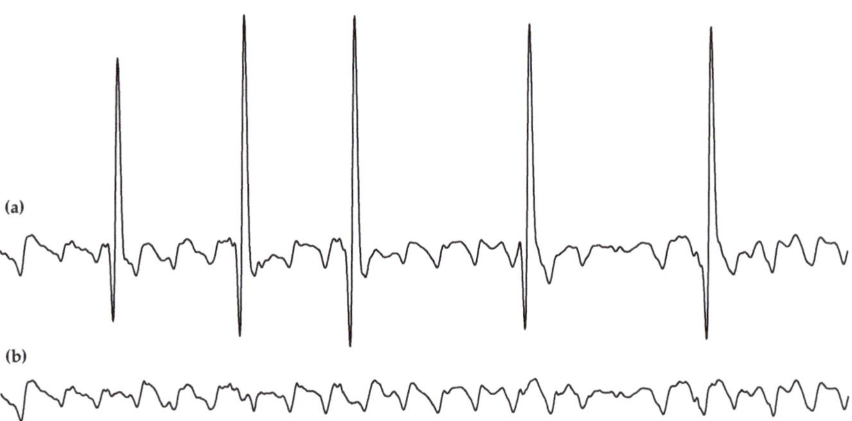

Figure 1. Example of a common (**a**) preprocessed ECG interval and (**b**) its *f*-waves extracted via a well-known QRST cancellation method [32].

2.3. Spectral Characterization of the *f*-Waves

Given the disparity in the ECG length acquired from the patients, the preprocessed recordings were segmented into 6 s intervals. For each patient, five consecutive intervals at most were considered, and the ECG-derived parameters from each one were averaged for unbiased subject-based analysis and classification. In this way, comparable values were obtained for all patients regardless of ECG signal duration, and slight variability noticed in *f*-waves owing to extracardiac noise [33] was minimized.

The power spectral density (PSD) of each 6 s ECG excerpt was estimated using the Welch Periodogram and is referred to as $\mathcal{W}(f)$. The computational parameters of this algorithm were selected to provide a spectral resolution of 0.1 Hz with a Hamming window 4000 points long and with 3000 points overlapping between adjacent windowed sections [23]. To serve as a reference, common spectral features previously proposed to anticipate CA outcome were then automatically computed from the *f*-wave segments. Thus, the DF was obtained as the frequency with the highest PSD amplitude [34,35], with this power value also considered in the study. Both parameters hereafter are referred to as f_0 and $\mathcal{W}(f_0)$, respectively. Making use of a 1 Hz bandwidth window centered on $2 \cdot f_0$, the first harmonic of the DF (f_1) and its spectral power ($\mathcal{W}(f_1)$) were also computed. The harmonic exponential decay (γ) was additionally estimated as a measure of the presence of harmonic components of the DF [25,26]. This parameter was defined as the logarithmic ratio of the spectral power between the DF and its first harmonic, i.e.,

$$\gamma = \ln\left(\frac{\mathcal{W}(f_0)}{\mathcal{W}(f_1)}\right). \tag{1}$$

The power of the DF and its harmonics has also been typically quantified through the well-known organization index ($\mathcal{O}$). It was defined as the ratio of the cumulative power under the DF and its first two harmonics and the area of the entire power spectrum between 3 and 25 Hz [36]. Both parameters γ and $\mathcal{O}$ consider only the power under the largest frequency components in the spectral distribution of the *f*-waves, not the shape and morphology of the harmonics, i.e., how the power is distributed around each frequency peak. Hence, in the present work, some pioneering parameters were considered for broader spectral characterization of the f-waves. More precisely, the shape of the DF and its harmonic components were separately analyzed by dividing the f-wave spectral distribution into two bands, as Figure 2 shows. The low-frequency (LF) band considered the spectral content around the DF, whereas the high-frequency (HF) band accounted for harmonic content. The cut-off frequency separating both bands was positioned approximately halfway between the DF and its first harmonic, that is, at three means of the DF value (f_0). For a global overview of the spectral distribution of the f-waves, the total frequency (TF) band ranging from 3 to 25 Hz was also analyzed.

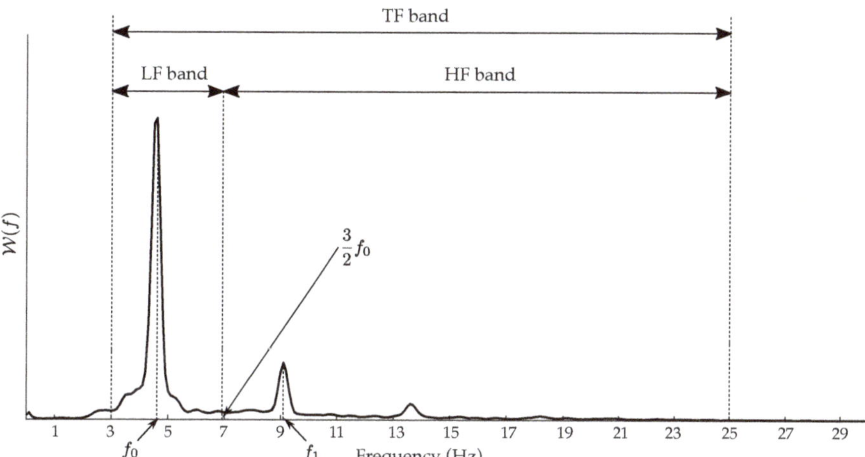

Figure 2. Spectral distribution of a common excerpt of f-waves divided into two frequency bands, i.e., low-frequency (LF) band and high-frequency (HF) band. Both form the total frequency (TF) band, ranging between 3 and 25 Hz.

The power distribution within each one of the three frequency bands was assessed through several entropy-based measures. Particularly, Wiener entropy, also known as spectral-flatness measure ($\mathcal{F}$), was firstly computed by dividing geometric and arithmetic means of the power contained by each band [37,38]; that is

$$\mathcal{F} = \frac{\sqrt[N]{\prod_{f=f_l}^{f_u} \mathcal{W}(f)}}{\frac{1}{N} \sum_{f=f_l}^{f_u} \mathcal{W}(f)},\tag{2}$$

where f_l and f_u are the lower and upper frequency band limits, respectively, and N is the total number of frequency samples. This index estimates the uniformity of signal-energy distribution in the frequency domain, so high values indicate more uniform or flat distributions, and low values are more peaky ones [38]. Another measure providing similar information is spectral entropy ($\mathcal{S}$), but this index was defined in a completely different way. In fact, $\mathcal{S}$ quantifies spectral complexity of a signal by computing the sparseness of its spectral distribution via Shannon entropy [39,40]. In brief, the PSD of the signal has to first be normalized by the total power of the frequency band of interest to obtain a probability function with unit area, i.e.,

$$\widetilde{\mathcal{W}}(f) = \frac{\mathcal{W}(f)}{\sum_{f=f_l}^{f_u} \mathcal{W}(f)}.\tag{3}$$

Then, $\mathcal{S}$ is estimated by computing the Shannon entropy from the resulting probability function, so that

$$\mathcal{S} = -\frac{1}{\ln(N)} \sum_{f=f_l}^{f_u} \widetilde{W}(f) \cdot \ln\left(\widetilde{W}(f)\right). \tag{4}$$

It should be noted that $\mathcal{S}$ is normalized by the highest possible value, i.e., $\ln(N)$, and thus ranges between 0 and 1. A high value of $\mathcal{S}$ implies a flat, uniform spectrum with a broad spectral content, whereas a low value implies a spectrum with all the power condensed into a single frequency bin, i.e., a less complex, more predictable signal [41].

A generalized version of $\mathcal{S}$ can be obtained by replacing the Shannon entropy with the Rényi entropy, thus obtaining Rényi spectral entropy ($\mathcal{R}$) as

$$\mathcal{R} = \frac{1}{\ln(N)} \cdot \frac{1}{1-\alpha} \ln\left(\sum_{f=f_l}^{f_u} \widetilde{W}(f)^\alpha\right), \tag{5}$$

where α ($\alpha \geq 0$ and $\alpha \neq 1$) is a bias parameter. Note that the Shannon entropy is an instance of Rényi entropy for $\alpha = 1$, and therefore, $\mathcal{R}$ might be a more sensitive tool than $\mathcal{S}$ to detect subtle changes in spectral distribution of a time series if α is appropriately chosen [41].

Finally, another measure of the spectral complexity of a signal is $\mathcal{C}_0$ complexity. This index is more robust to conditions of nonlinearity and non-stationarity than previous ones [42]. For its computation, the normalized PSD of the f-waves was modified to only preserve its most irregular part, i.e.,

$$\widehat{W}(f) = \begin{cases} \widetilde{W}(f), & if\ \widetilde{W}(f) \leq T, \\ 0, & if\ \widetilde{W}(f) > T, \end{cases} \tag{6}$$

with T being a threshold computed from the mean spectral power, so that

$$T = \frac{2}{N} \sum_{f=f_l}^{f_u} \widetilde{W}(f). \tag{7}$$

Then, the $\mathcal{C}_0$ complexity is computed as the power ratio of the irregular part to the original signal [42], i.e.,

$$\mathcal{C}_0 = \sum_{f=f_l}^{f_u} \frac{\widehat{W}(f)}{\widetilde{W}(f)}, \tag{8}$$

resulting in a real number between 0 and 1, so that the greater the predominance of the irregular part of the signal, the higher the value of $\mathcal{C}_0$.

2.4. Statistical Analysis and Classification Performance

Normality and homoscedasticity of continuous variables were firstly assessed using Lilliefors and Levene's tests, respectively. When both conditions were met, a parametric Student's t-test was used to measure statistical differences between the two groups of patients, i.e., those who maintained SR and those who relapsed to AF after a follow-up of 9 months. A non-parametric Mann–Whitney U-test was used for the same purpose when data distributions were non-normal but homoscedastic. Since most parameters showed normal and homoscedastic distributions, values of all the features are summarized along the manuscript in terms of mean $\pm$ standard deviation. Regarding categorical variables, they are reported as number and percentage, and were compared using a Fisher exact test. In all cases, a value of significance $p < 0.05$ was considered as statistically significant.

On the other hand, the classification performance of each single feature was evaluated through a repeated, patient-based, 10-fold cross-validation approach [43]. Precisely, in each cross-validation procedure, the data were first partitioned into 10 equally sized folds. Subsequently, 10 iterations of training and validation were performed, such that within each one a different fold of the data was held out for validation while the remaining nine folds were used for learning. The data were stratified by ensuring that each fold was a good representative of the whole. Linear discriminant analysis (LDA) was used to train a prediction model in each iteration. This analysis assumes that different classes generate data based on Gaussian distributions, such that training an LDA model involves finding the parameters for a Gaussian distribution for each class. In fact, the procedure searches for a projection hyperplane of the observations for which the variance of each class is minimized and the distance between the means of the classes is maximized [44].

The classification results for each 10-fold cross-validation procedure are summarized by means of a receiver operating characteristic (ROC) curve computed on the classification scores provided by the obtained LDA models. This plot is the result of plotting the fraction of true positives out of total positives (sensitivity) against the fraction of false positives out of total negatives (1-specificity) at various thresholds. Sensitivity (Se) was considered to be the rate of patients who were correctly classified as relapsing to AF, whereas specificity (Sp) was the percentage of patients properly identified as maintaining SR. An optimal threshold for separating both groups was selected to provide the best balance between Se and Sp, although in this way, the highest percentage of patients correctly classified, i.e., accuracy (Acc), could not be achieved [45]. The area under the ROC curve (AUC) was also obtained as an aggregate performance measure across all possible thresholds [45]. To provide additional information about the proportions of positive and negative samples that were true positives and true negatives, the positive predictive value (PPV) and negative predictive value (NPV) were also computed. The described validation process was repeated 100 times to obtain general and unbiased classification outcomes [43]. The data were reshuffled and re-stratified before each 10-fold cross-validation approach, and Se, Sp, Acc, AUC, PPV, and NPV values were averaged for the 100 cycles.

Finally, to explore complementary information among single features and to improve prediction of CA outcome after a follow-up of 9 months, a multivariate analysis was conducted. In this case, a LDA was also used to build prediction models based on linear combinations of those features automatically selected by making use of a forward sequential selection technique. In this analysis, features were sequentially added to an empty candidate set until the addition of further features did not decrease the criterion function, i.e., the prediction error [46]. That error was assessed inside repeated cross-validation loops to avoid any bias in feature selection [43]. Hence, variables that were selected more frequently for the models were used last to build several LDA-based prediction algorithms and were evaluated as single features, i.e., by running the patient-based 10-fold cross-validation 100 times. The classification improvement achieved by these models regarding single features was statistically evaluated using an asymptotic McNemar's test. To compare the accuracies of two classification models, the algorithm first compared their predicted labels against the true labels and then detected whether the difference between the misclassification rates was statistically significant [47].

3. Results

After a follow-up of 9 months, 103 patients maintained SR, and the remaining 48 relapsed to AF. This implies that the CA procedure was unsuccessful in the mid-term for 31.79% of the patients, which is consistent with current AF recurrence statistics [9,11]. The baseline clinical characteristics of both groups are provided in Table 1. As can be seen, none of the features collected for the patients, i.e., gender, age, AF duration before the CA procedure, body mass index, and left atrium diameter, presented statistically significant differences between the patients who maintained SR and those who relapsed to AF.

Table 1. Baseline clinical characteristics of the population enrolled in the study with their corresponding statistical significance to distinguish between the two groups of patients.

Clinical Feature	Rhythm after Follow-Up		p-Value
	SR	AF	
Number of patients (%)	103 (68.21%)	48 (31.79%)	−
Male (%)	79 (76.70%)	37 (77.08%)	1.000
Age (years)	59.37 ± 12.24	57.23 ± 12.82	0.326
With AF <1 year (%)	6 (5.83%)	6 (12.50%)	0.198
With AF 1–3 years (%)	70 (67.96%)	31 (64.58%)	0.713
With AF >3 years (%)	27 (26.21%)	11 (22.92%)	0.841
Body mass index (kg/m^2)	27.79 ± 3.55	29.06 ± 4.86	0.073
Left atrium diameter (mm)	44.11 ± 5.70	45.65 ± 5.32	0.117

Regarding the wide range of features considered in the present work to characterize f-wave spectral distribution, Table 2 shows the values obtained for the two group of patients and the corresponding statistical significance (p-value). Among the parameters commonly proposed in previous works, f_0, f_1, and γ provided statistically significant differences between patients who maintained SR and relapsed to AF, with the former having lower values for the three indices. Similarly, the four proposed entropy- and complexity-based features also provided statistically significant differences between both groups of patients when the TF band covering the whole f-wave distribution between 3 and 25 Hz was considered. In the case of the LF band, mainly containing the DF component, $\mathcal{F}$ and $\mathcal{C}_0$ also provided statistically significant differences between the two groups of patients, and $\mathcal{R}$ reported a tendency close to being significant. On the contrary, no relevant differences between the groups of patients were noticed in any of the four indices in the frequency band covering the harmonic content, i.e., the HF band. It should be noted that several values of α between 0.1 and 2 were tested to compute the Rényi spectral entropy, but the best results (i.e., those presented in Table 2) were obtained for $\alpha = 0.1$.

The classification performance of the single features reporting statistically significant values of $p < 0.05$ is displayed in Table 3. Most indices presented values of Se, Sp, Acc, and AUC greater than 60%, but only γ and $\mathcal{C}_{0TF}$ exhibited performance metrics higher than 70%. Both parameters reported similar classification performances, which were statistically better than that provided by f_0 for all the conducted validation cycles according to McNemar's test. These two variables also provided the highest values of PPV and NPV, about 52% and 83%, thus improving the DF results by more than 15 and 10%, respectively. Similarly, the classification results reported by Wiener entropy and Rényi spectral entropy for the TF band (i.e., $\mathcal{F}_{TF}$ and $\mathcal{R}_{TF}$) were also statistically superior to that of f_0, but in this case, values of Acc and AUC of about 67%, values of PPV of about 47%, and values of NPV of about 81% were obtained. Contrarily, no statistically significant differences in the classification performance was noticed between f_0 and the entropy-based indices computed on the LF band, i.e., $\mathcal{F}_{LF}$ and $\mathcal{C}_{0LF}$.

On the other hand, multivariate analysis showed that the parameters most frequently selected for the LDA-based models built throughout the validation cycles were γ and $\mathcal{F}_{TF}$. Nonetheless, instead of the last parameter, $\mathcal{C}_{0TF}$ and $\mathcal{R}_{TF}$ were sometimes selected, along with the index γ. The classification performance of these three LDA-based models is presented in the first rows of Table 4. As can be observed, they were very similar for the three cases, reaching values of Acc and AUC of about 75% and values of PPV and NPV of about 58% and 86%, respectively. Notably, the three prediction models obtained classification results statistically better than those of the included single features (i.e., γ, $\mathcal{F}_{TF}$, $\mathcal{C}_{0TF}$, and $\mathcal{R}_{TF}$) and better than f_0 for most validation cycles according to McNemar's test. In fact, improvements in values of Acc, AUC, PPV, and NPV greater than 5% were obtained by the LDA-based models compared to those of the included single features and by about 20% compared to those of the DF. Finally, the inclusion of a third variable to these prediction models did not improve the classification results. For instance, the last three rows of Table 4 display how Acc and AUC slightly decreased when the models were complemented with another feature, even when it was related to the DF component (such as $\mathcal{R}_{LF}$) and was sometimes chosen by the automated feature selection algorithm. Along a similar line, the inclusion of any clinical and/or echocardiographic variables presented in Table 1 improved classification performance of the prediction models.

Table 2. Mean and standard deviation values for the analyzed metrics from the two groups of patients and their corresponding statistical significance (*p*-value).

Feature	Rhythm after the Follow-Up		*p*-Value
	SR	**AF**	
f_0 (Hz)	5.69 ± 1.12	6.14 ± 0.99	0.009
$\mathcal{W}(f_0)$ (mV^2)	0.00051 ± 0.00091	0.00060 ± 0.00092	0.059
f_1 (Hz)	11.34 ± 2.25	12.28 ± 1.98	0.008
$\mathcal{W}(f_1)$ (mV^2)	0.00013 ± 0.00041	0.00005 ± 0.00009	0.754
γ	2.20 ± 0.77	2.80 ± 0.57	<0.001
$\mathcal{O}$	0.52 ± 0.15	0.53 ± 0.12	0.488
$\mathcal{F}_{LF}$	0.48 ± 0.15	0.43 ± 0.12	0.017
$\mathcal{S}_{LF}$	0.82 ± 0.08	0.81 ± 0.06	0.224
$\mathcal{R}_{LF}$	0.980 ± 0.010	0.978 ± 0.008	0.065
$\mathcal{C}_{0LF}$	0.45 ± 0.14	0.41 ± 0.11	0.045
$\mathcal{F}_{HF}$	0.49 ± 0.13	0.50 ± 0.10	0.865
$\mathcal{S}_{HF}$	0.86 ± 0.07	0.87 ± 0.05	0.828
$\mathcal{R}_{HF}$	0.985 ± 0.007	0.986 ± 0.005	0.967
$\mathcal{C}_{0HF}$	0.48 ± 0.10	0.49 ± 0.08	0.550
$\mathcal{F}_{TF}$	0.28 ± 0.10	0.22 ± 0.06	<0.001
$\mathcal{S}_{TF}$	0.76 ± 0.07	0.73 ± 0.06	0.008
$\mathcal{R}_{TF}$	0.974 ± 0.008	0.971 ± 0.006	<0.001
$\mathcal{C}_{0TF}$	0.32 ± 0.07	0.28 ± 0.05	<0.001

Table 3. Classification between patients who relapsed to AF and maintained SR during the follow-up based on the most-predictive single parameters.

Feature	Se (%)	Sp (%)	Acc (%)	AUC	PPV (%)	NPV (%)
f_0	56.25	56.42	56.36	0.617	37.56	73.45
f_1	56.33	56.68	56.57	0.620	37.73	73.58
γ	69.29	69.18	69.22	0.728	51.17	82.86
$\mathcal{F}_{LF}$	60.27	60.33	60.31	0.606	41.45	76.52
$\mathcal{C}_{0LF}$	56.54	56.71	56.66	0.587	37.84	73.68
$\mathcal{F}_{TF}$	66.13	66.22	66.19	0.676	47.71	80.75
$\mathcal{S}_{TF}$	58.83	59.28	59.14	0.623	40.24	75.55
$\mathcal{R}_{TF}$	66.10	66.18	66.16	0.675	47.67	80.73
$\mathcal{C}_{0TF}$	70.83	70.91	70.89	0.703	53.16	83.92

Table 4. Classification between patients who relapsed to AF and maintained SR during the follow-up provided by the prediction models obtained from multivariable linear discriminant analysis.

Features in the Model	Se (%)	Sp (%)	Acc (%)	AUC	PPV (%)	NPV (%)
γ and $\mathcal{F}_{TF}$	72.75	75.14	74.38	0.758	57.69	85.54
γ and $\mathcal{R}_{TF}$	72.77	74.89	74.22	0.748	57.46	85.51
γ and $\mathcal{C}_{0TF}$	73.04	74.66	74.15	0.759	57.33	85.60
γ, $\mathcal{F}_{TF}$, and $\mathcal{C}_{0TF}$	72.15	74.56	73.79	0.751	56.93	85.17
γ, $\mathcal{F}_{LF}$, and $\mathcal{R}_{TF}$	73.90	73.12	73.36	0.743	56.16	85.74
γ, $\mathcal{R}_{TF}$, and $\mathcal{C}_{0TF}$	70.31	74.90	73.44	0.753	56.63	84.41

4. Discussion

Despite its limited success rate in the medium/long term, CA remains a commonly used treatment for persistent AF [9]. Hence, clinical interest in the preoperative prediction of the outcome of this intervention has substantially grown in the last years [1], since it might anticipate those patients with a high probability of early AF recurrence [48]. Keeping in mind that CA often involves long-lasting procedures and high costs and risks for patients unable to maintain SR for prolonged periods of time, the selection of optimal candidates for this treatment would have interesting benefits. In this respect, not only could the huge costs associated with AF treatment be reduced, but the risks related to the CA procedure would be limited, tailored approaches could be enabled, and the number of repeated interventions could be minimized in many patients, among other benefits [13].

So far, some clinical predictors of arrhythmia recurrence after CA have been explored, such as AF duration, anatomical characteristics of the left atrium, presence of concomitant diseases, etc. However, they have only provided controversial results with limited predictive ability, which is usually lower in persistent rather than paroxysmal AF patients [14,49,50]. Accordingly, in the present study, none of the evaluated clinical variables were relevant in the prediction of patients who relapsed to AF versus those who maintained SR after a follow-up of 9 months. On the other hand, several previous works have introduced the development of predictors based on quantitative analysis of atrial electrical conduction. Invasive studies have shown that the DF and its inverse (i.e., the AFCL) have a moderate predictive capacity of midterm SR maintenance after CA [18–21,51,52]. However, these predictors require invasive recordings of the atrial electrical activity by means of electrograms acquired during the procedure, thus entailing an unnecessary risk in those patients for whom the CA procedure will not be successful.

Subsequently, non-invasive studies have also proven the predictive potential of manual measurement of the AFCL on surface ECGs, yielding a higher preoperative value for patients with higher probability of long-term SR maintenance after CA [15]. More recently, once f-waves were extracted from the surface ECG, the automated measurement of the DF has also shown a reasonable ability to anticipate CA outcome in persistent AF patients, associating lower values of frequency with a lower probability of AF recurrence in the medium/long term [34,35]. The results obtained by the DF in the present work are consistent with that tendency (see Table 2), and since this index has been directly linked to the degree of electrical remodeling presented by the atria [34], it can be considered a good reference for comparison with other predictors. The frequency of the DF's first harmonic (f_1) also presented the same trend in the results for both groups of patients, exhibiting predictive ability similar to that of the DF with Acc, AUC, PPV, and NPV values about 57%, 62%, 38%, and 74%, respectively (see Table 2).

In contrast to these results, the power peak both for the DF and its first harmonic did not reveal statistically significant differences between patients who relapsed to AF and those who maintained SR after the follow-up. Along the same line, the organization index $\mathcal{O}$ was also unable to discern between the two groups of patients, thus suggesting that the power globally concentrated around the DF and its first harmonic is not relevant in the prediction of CA outcome. Similar results have also been reported in other works [23,36]. However, the power ratio of the first harmonic to the DF, i.e., the index γ, proved to be one of the best single predictors, with values of Acc and AUC of about 70%, and values of PPV and NPV of about 51% and 83%, respectively (see Table 2). A similar finding was previously outlined in a previous work, where γ was the only spectral parameter able to report a statistically relevant significance on a limited dataset of 22 persistent AF patients under radio-frequency CA [23]. In the present work, that outcome has been corroborated on a much wider database of 151 patients. According to the definition of γ, low values indicate the presence of strong harmonics of the DF, and the obtained results hence suggest that the higher the harmonic content of the DF, the lower the probability of AF recurrence after CA. This finding is confirmed in Figure 3, which shows mean spectral distributions for all the 6 s ECG segments from the patients who relapsed to AF compared to those who maintained

SR after the follow-up once they were aligned with respect to the DF. As can be seen, on average, the patients who maintained SR exhibited greater harmonics of the DF than those who relapsed to AF. Interestingly, the presence of larger harmonics has also been associated with increased probability of AF termination during CA [18] and with medium/long-term maintenance of SR after diverse AF therapies, including pharmacological cardioversion [26] and internal [24] and external electrical cardioversion [25].

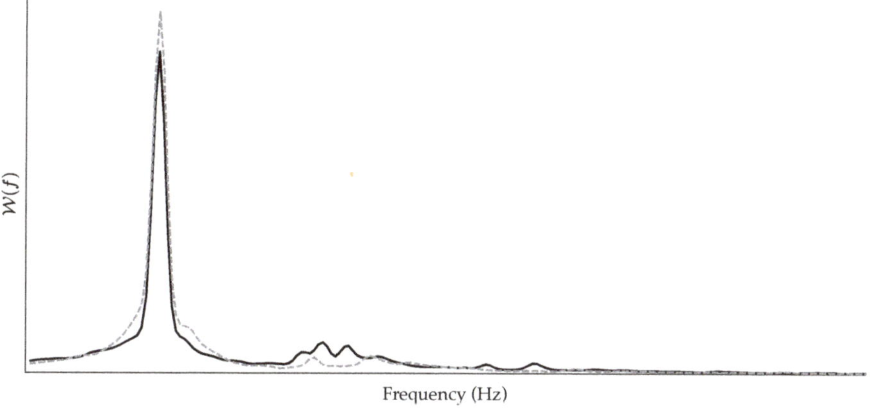

Figure 3. Averaged spectral distributions for all the 6 s ECG segments from patients who relapsed to AF (dashed, gray line) and those who maintained SR after the follow-up (solid, black line) once they were aligned with respect to the DF.

To further delve into the analysis of the f-wave harmonic spectral structure, four entropy-based indices were computed for the first time from three frequency bands. All intended to quantify how power was distributed along the spectra and then to separately and globally characterize the shape of the DF and its harmonic content. However, each index presented a different mathematical definition and were thus sensitive to different shades in a time series [53]. The Wiener entropy $\mathcal{F}$ has been widely studied to predict the results of electrical defibrillation in ventricular fibrillation [37] and is computed by dividing geometric and arithmetic means of the signal spectrum. In a completely different way, spectral entropy $\mathcal{S}$ and Rényi spectral entropy $\mathcal{R}$ treat the normalized power distribution of the f-waves as a probability distribution and calculate Shannon entropy and Rényi entropy, respectively. The first index has already been proposed to discern between patients with persistent and permanent AF [39] and to anticipate spontaneous termination of paroxysmal AF episodes [40]. It should be noted that Rényi entropy is a generalization of Shannon entropy and is thus able to provide additional information [41]. In fact, in the present work, $\mathcal{R}$ provided notably better results than $\mathcal{S}$ when computed on the LF and TF bands. In the former case, although both indices did not provide statistically significant differences between patients who relapsed to AF versus those who maintained SR after the follow-up, the index $\mathcal{R}_{LF}$ reported a nearly significant difference ($p = 0.065$, see Table 2). Regarding the TF band, $\mathcal{R}_{TF}$ provided statistically significant improvements of about 5% in values of Acc, AUC, PPV, and NPV compared to $\mathcal{S}_{TF}$ according to McNemar's test. Finally, the $\mathcal{C}_0$ complexity divides spectral distribution of a time series to estimate its regular and irregular parts and then obtains an organization estimate highly robust to noise and non-stationary artifacts [42].

Despite slight differences in their results, the indices $\mathcal{F}$, $\mathcal{R}$, and $\mathcal{C}_0$ reported similar general trends on the three analyzed frequency bands of the f-waves. In the case of the LF band, negligible differences on the edge of being statistically significant were noticed by the three parameters (see Table 2). This finding suggests that the DF component presented a kindred shape for the patients who relapsed to AF versus those who maintained SR after the follow-up. However, the higher values of entropy and complexity noticed in the patients who maintained SR, along with the trend to higher power peaks ($\mathcal{W}(f_0)$) in those who relapsed to AF, point toward the presence of a mildly larger and more-peaked DF component when the probability of AF recurrence after CA increased. This suggestion is supported by Figure 3, where a slightly larger and wider dominant peak can be seen on average for the patients who relapsed to AF. Regarding the HF band, no differences were noticed by any entropy-based parameter, suggesting that both groups of patients presented a similarly well-defined structure of harmonic components of the DF, regardless of the power contained by them. This assumption can also be visually corroborated in Figure 3. Nonetheless, it should be remarked that the indices $\mathcal{F}$, $\mathcal{R}$, and $\mathcal{C}_0$ consider the power distribution along the spectrum, but not the absolute level of power [53], which was highly variable within each group of patients according to the values provided by the variable $\mathcal{W}(f_1)$ in Table 2.

These results for the LF and HF bands contrast with those obtained when the TF band, ranging from 3 to 25 Hz, was analyzed. In this case, the parameters $\mathcal{F}_{TF}$, $\mathcal{R}_{TF}$, and $\mathcal{C}_{0TF}$ found highly marked, statistically significant differences and notably high classification performance between the patients who relapsed to AF and those who maintained SR after the follow-up. Indeed, they presented statistically significant improvements of about 10% in values of Se, Sp, Acc, AUC, PPV, and NPV regarding the DF and their computation on the LF band, i.e., $\mathcal{F}_{LF}$, $\mathcal{R}_{LF}$, and $\mathcal{C}_{0LF}$ (see Table 3). Moreover, along with the index γ, the variable $\mathcal{C}_{0TF}$ reported the highest performance metrics, with values about 70%. This wide disparity among frequency bands and the good results of the index γ suggest that the most useful information for preoperative prediction of CA outcome lies in the relation between the DF and its harmonic structure and not in the single information individually provided by each one. In this respect, it is interesting to highlight that the parameters $\mathcal{F}_{TF}$, $\mathcal{R}_{TF}$, and $\mathcal{C}_{0TF}$ globally evaluated the whole spectrum of the f-waves before taking into consideration both the DF and its harmonic content. The greater entropy and complexity values observed in the patients who maintained SR during the follow-up could hence be explained by the fact that they presented larger harmonics and a proportionally smaller peak DF than those who relapsed to AF. This assumption can be visually corroborated in Figure 3 on average for the two groups of patients.

The same idea of the presence of a larger DF component with low-amplitude harmonics in the patients who relapsed to AF also underlies the values reported by the variable γ. However, the relation of this index with $\mathcal{F}_{TF}$, $\mathcal{R}_{TF}$, and $\mathcal{C}_{0TF}$ was not as strong as initially expected. In fact, parametric analysis reported notably low correlation values, which were on the edge of being statistically significant, of 16.36% ($p = 0.045$), 11.96% ($p = 0.144$), and 17.96% ($p = 0.027$) for the pairs γ and $\mathcal{F}_{TF}$, γ and $\mathcal{R}_{TF}$, and γ and $\mathcal{C}_{0TF}$, respectively. Moreover, the conducted multivariable analysis also provided that the combination of these pairs of parameters in LDA-based prediction models obtained statistically significant improvements of about 5% in values of Se, Sp, Acc, and AUC regarding the single variables (see Tables 2 and 3), thus achieving the best performance metrics of about 75%. Similarly, PPV and NPV also experienced a statistically significant increase of about 5% regarding the single parameters combined in these models and of up to 20% in comparison with the DF. As a consequence, the index γ and the entropy-based parameters computed on the TF band could contain complementary information, and therefore, joint analysis of both the power ratio between the DF and its first harmonic and the global distribution of the power along the spectrum seems to play a key role to improve the preoperative prediction of CA outcome in persistent AF patients.

Extensive ablation based on linear lesions or complex fractionated electrogram (CFE) ablation in addition to PVI has been widely proposed in the cientific literature and widely used in clinical practice. However, large, multicenter and prospective studies comparing these CA strategies with PVI alone have failed to provide any additional benefit, but prolong the duration of the procedure [54,55]. Thereby, all patients enrolled in the present study were treated with PVI alone. The adoption of this CA strategy provides the fairest comparison between patients. In fact, all subjects received the same atrial substrate modification, thus avoiding a potential bias in the study and allowing the analyzed preoperative predictors to focus on quantifying main differences between atrial electrophysiological features of the patients with high and low risk of midterm AF recurrence. Contrarily, when a tailored CA protocol is applied to each patient by PVI plus a variable number of linear lesions or targeted CFEs, it is reasonable to think that the preoperative predictors could be impacted by the different degree of modification provoked in the atrial substrate of each patient. However, to the best of our knowledge, this hypothesis has still not been corroborated. Some previous works have considered individualized protocols for each patient, but disaggregated data on each kind of intervention (i.e., PVI alone, PVI plus linear lesions, and PVI plus CFE ablation) have not been provided to date [15,34–36,52,56]. These studies have reported a high disparity in the predictive power and cut-off point for some well-established parameters, such as the DF, and the use of different CA protocols could partially explain this finding. Clearly, the proposed predictors based on the f-wave harmonic spectral structure could be similarly impacted by the use of tailored CA approaches, but this aspect will have to be addressed in further studies.

Finally, some limitations of the study merit comment. Following current clinical guidelines [7], follow-up of patients after CA was mainly based on standard ECG and 24 h Holter monitoring at different scheduled visits and, in case of symptoms, additional exploration and ECG recordings in the emergency room. However, since continuous ECG monitoring for the whole follow-up was not used, some asymptomatic, non-sustained AF episodes may have been overlooked, and the number of patients with AF recurrence may have been underestimated. Moreover, a follow up of 9 months could be considered a short period of time to assess CA outcome; however, this midterm AF recurrence prediction is still clinically interesting to select optimal candidates for the intervention. In fact, many previous works have addressed CA outcome predictions at similar or even shorter periods, such as [15–17,34–36]. Nonetheless, a longer follow-up period will be considered in future work, since visits every 12 months are scheduled for all the patients. On the other hand, the sole analysis of the V1 lead precluded specific information from the electrical activity registered at many atrial regions. Although this lead has proved to be the best to reflect global activation of the atria and has, moreover, been widely analyzed in the scientific literature [16,17,23,57], there is recent evidence that the study of spatial variability of ECG-based parameters could be helpful to improve CA outcome prediction [58]. Hence, a multi-lead extension of f-wave harmonic structure analysis will be conducted in the future. Lastly, although a comparable [34–36] or much larger number of patients [23,52,56,58] than in previous works also dealing with CA outcome prediction has been analyzed in the present study, the database was retrospective and only came from two centers. Hence, in the future, the conclusions will be corroborated in wider datasets prospectively collected from a higher number of hospitals. A broader analysis will also be performed, considering more predictors, since the results provided by the indices of this study had a limited PPV value in comparison to NPV.

5. Conclusions

The present work has conducted a pioneering analysis of the f-wave harmonic spectral structure to improve preoperative prediction of CA outcome in persistent AF patients. The results show that the relation between the DF and its harmonic content contains more-relevant information for prediction than separate analysis of each frequency component. While the DF and its harmonic structure individually presented similar global shapes both for patients who relapsed to AF and those who maintained SR after the follow-up, the power ratio between both components had the best discriminant ability. Indeed, the presence of larger harmonics and a proportionally smaller DF peak was strongly associated with a decreased probability of AF recurrence after CA. Moreover, analysis of the global distribution of the f-wave power along the spectrum through diverse entropy-based indices, jointly considering both the DF and its harmonic content, also revealed complementary information with respect to their power ratio, thus significantly improving the preoperative prediction of CA outcome.

Author Contributions: Conceptualization, P.E., J.J.R. and R.A.; methodology, P.E., J.J.R. and R.A.; software, P.E., J.R. and M.G.; validation, J.J.R. and R.A.; resources, M.A.A., V.M.H. and S.C.; data curation, P.E., J.R., M.G. and R.A.; writing—original draft preparation, P.E.; writing—review and editing, J.R., M.G., M.A.A., V.M.H., S.C., J.J.R. and R.A. All authors have read and agreed to the published version of the manuscript.

Funding: This research received financial support from public grants PID2021-00X128525-IV0, PID2021-123804OB-I00, and TED2021-130935B-I00 of the Spanish Government 10.13039/501100011033 jointly with the European Regional Development Fund (EU), SBPLY/17/180501/000411 and SB-PLY/21/180501/000186 from Junta de Comunidades de Castilla-La Mancha, and AICO/2021/286 from Generalitat Valenciana. Moreover, Pilar Escribano holds a predoctoral scholarship 2020-PREDUCLM-15540, which is co-financed by the operating program of the European Social Fund (ESF) 2014–2020 of Castilla-La Mancha.

Institutional Review Board Statement: The study was conducted according to the guidelines of the Declaration of Helsinki, complied with Law 14/2007, 3rd of July, on Biomedical Research and other Spanish regulations, and was approved by the Ethical Review Board of University Hospitals of Toledo and Albacete, Spain (protocol code 5064, 1 December 2020).

Informed Consent Statement: Informed consent was received from all the subjects participating in the present research. All acquired data were anonymized before processing.

Data Availability Statement: The data supporting reported results and presented in this study are available on request from the corresponding author.

Conflicts of Interest: The authors declare no conflict of interest.

References

1. Platonov, P.G.; Corino, V.D.A. A Clinical Perspective on Atrial Fibrillation. In *Atrial Fibrillation from an Engineering Perspective*; Springer International Publishing: Berlin, Germany, 2018; pp. 1–24. [CrossRef]
2. Lippi, G.; Sanchis-Gomar, F.; Cervellin, G. Global epidemiology of atrial fibrillation: An increasing epidemic and public health challenge. *Int. J. Stroke* **2021**, *16*, 217–221. [CrossRef] [PubMed]
3. Zoni-Berisso, M.; Lercari, F.; Carazza, T.; Domenicucci, S. Epidemiology of atrial fibrillation: European perspective. *Clin. Epidemiol.* **2014**, *6*, 213–220. [CrossRef] [PubMed]
4. Schotten, U.; Dobrev, D.; Platonov, P.G.; Kottkamp, H.; Hindricks, G. Current controversies in determining the main mechanisms of atrial fibrillation. *J. Intern. Med.* **2016**, *279*, 428–438. [CrossRef] [PubMed]
5. Rienstra, M.; Lubitz, S.A.; Mahida, S.; Magnani, J.W.; Fontes, J.D.; Sinner, M.F.; Van Gelder, I.C.; Ellinor, P.T.; Benjamin, E.J. Symptoms and functional status of patients with atrial fibrillation: state of the art and future research opportunities. *Circulation* **2012**, *125*, 2933–2943. [CrossRef]
6. Warmus, P.; Niedziela, N.; Huć, M.; Wierzbicki, K.; Adamczyk-Sowa, M. Assessment of the manifestations of atrial fibrillation in patients with acute cerebral stroke—A single-center study based on 998 patients. *Neurol. Res.* **2020**, *42*, 471–476. [CrossRef]
7. Hindricks, G.; Potpara, T.; Dagres, N.; Arbelo, E.; Bax, J.J.; Blomström-Lundqvist, C.; Boriani, G.; Castella, M.; Dan, G.A.; Dilaveris, P.E.; et al. 2020 ESC Guidelines for the diagnosis and management of atrial fibrillation developed in collaboration with the European Association for Cardio-Thoracic Surgery (EACTS): The Task Force for the diagnosis and management of atrial fibrillation of the European Society of Cardiology (ESC) Developed with the special contribution of the European Heart Rhythm Association (EHRA) of the ESC. *Eur. Heart J.* **2021**, *42*, 373–498. [CrossRef]
8. Nattel, S.; Guasch, E.; Savelieva, I.; Cosio, F.G.; Valverde, I.; Halperin, J.L.; Conroy, J.M.; Al-Khatib, S.M.; Hess, P.L.; Kirchhof, P.; et al. Early management of atrial fibrillation to prevent cardiovascular complications. *Eur. Heart J.* **2014**, *35*, 1448–1456. [CrossRef]
9. Schmidt, B.; Brugada, J.; Arbelo, E.; Laroche, C.; Bayramova, S.; Bertini, M.; Letsas, K.P.; Pison, L.; Romanov, A.; Scherr, D.; et al. Ablation strategies for different types of atrial fibrillation in Europe: Results of the ESC-EORP EHRA Atrial Fibrillation Ablation Long-Term registry. *Europace* **2020**, *22*, 558–566. [CrossRef]
10. Calkins, H.; Reynolds, M.R.; Spector, P.; Sondhi, M.; Xu, Y.; Martin, A.; Williams, C.J.; Sledge, I. Treatment of atrial fibrillation with antiarrhythmic drugs or radiofrequency ablation: Two systematic literature reviews and meta-analyses. *Circ. Arrhythm. Electrophysiol.* **2009**, *2*, 349–361. [CrossRef]
11. Hesselson, A.B. Catheter Ablation in the Treatment of Atrial Fibrillation. *Int. J. Angiol.* **2020**, *29*, 108–112. [CrossRef]
12. Calkins, H.; Hindricks, G.; Cappato, R.; Kim, Y.H.; Saad, E.B.; Aguinaga, L.; Akar, J.G.; Badhwar, V.; Brugada, J.; Camm, J.; et al. 2017 HRS/EHRA/ECAS/APHRS/SOLAECE expert consensus statement on catheter and surgical ablation of atrial fibrillation: Executive summary. *J. Arrhythm.* **2017**, *33*, 369–409. [CrossRef] [PubMed]

13. Walsh, K.; Marchlinski, F. Catheter ablation for atrial fibrillation: Current patient selection and outcomes. *Expert Rev. Cardiovasc. Ther.* **2018**, *16*, 679–692. [CrossRef] [PubMed]

14. Balk, E.M.; Garlitski, A.C.; Alsheikh-Ali, A.A.; Terasawa, T.; Chung, M.; Ip, S. Predictors of atrial fibrillation recurrence after radiofrequency catheter ablation: A systematic review. *J. Cardiovasc. Electrophysiol.* **2010**, *21*, 1208–1216. [CrossRef] [PubMed]

15. Matsuo, S.; Lellouche, N.; Wright, M.; Bevilacqua, M.; Knecht, S.; Nault, I.; Lim, K.T.; Arantes, L.; O'Neill, M.D.; Platonov, P.G.; et al. Clinical predictors of termination and clinical outcome of catheter ablation for persistent atrial fibrillation. *J. Am. Coll. Cardiol.* **2009**, *54*, 788–795. [CrossRef]

16. Cheng, Z.; Deng, H.; Cheng, K.; Chen, T.; Gao, P.; Yu, M.; Fang, Q. The amplitude of fibrillatory waves on leads aVF and V1 predicting the recurrence of persistent atrial fibrillation patients who underwent catheter ablation. *Ann. Noninvasive Electrocardiol.* **2013**, *18*, 352–358. [CrossRef]

17. Nault, I.; Lellouche, N.; Matsuo, S.; Knecht, S.; Wright, M.; Lim, K.T.; Sacher, F.; Platonov, P.; Deplagne, A.; Bordachar, P.; et al. Clinical value of fibrillatory wave amplitude on surface ECG in patients with persistent atrial fibrillation. *J. Interv. Card. Electrophysiol.* **2009**, *26*, 11–19. [CrossRef]

18. Takahashi, Y.; Sanders, P.; Jaïs, P.; Hocini, M.; Dubois, R.; Rotter, M.; Rostock, T.; Nalliah, C.J.; Sacher, F.; Clémenty, J.; et al. Organization of frequency spectra of atrial fibrillation: Relevance to radiofrequency catheter ablation. *J. Cardiovasc. Electrophysiol.* **2006**, *17*, 382–388. [CrossRef]

19. Yoshida, K.; Ulfarsson, M.; Tada, H.; Chugh, A.; Good, E.; Kuhne, M.; Crawford, T.; Sarrazin, J.F.; Chalfoun, N.; Wells, D.; et al. Complex electrograms within the coronary sinus: Time- and frequency-domain characteristics, effects of antral pulmonary vein isolation, and relationship to clinical outcome in patients with paroxysmal and persistent atrial fibrillation. *J. Cardiovasc. Electrophysiol.* **2008**, *19*, 1017–1023. [CrossRef]

20. Atienza, F.; Almendral, J.; Jalife, J.; Zlochiver, S.; Ploutz-Snyder, R.; Torrecilla, E.G.; Arenal, A.; Kalifa, J.; Fernández-Avilés, F.; Berenfeld, O. Real-time dominant frequency mapping and ablation of dominant frequency sites in atrial fibrillation with left-to-right frequency gradients predicts long-term maintenance of sinus rhythm. *Heart Rhythm* **2009**, *6*, 33–40. [CrossRef]

21. Yoshida, K.; Chugh, A.; Good, E.; Crawford, T.; Myles, J.; Veerareddy, S.; Billakanty, S.; Wong, W.S.; Ebinger, M.; Pelosi, F.; et al. A critical decrease in dominant frequency and clinical outcome after catheter ablation of persistent atrial fibrillation. *Heart Rhythm* **2010**, *7*, 295–302. [CrossRef]

22. Sörnmo, L.; Alcaraz, R.; Laguna, P.; Rieta, J.J. Characterization of f Waves. In *Atrial Fibrillation from an Engineering Perspective*; Springer International Publishing: Berlin, Germany, 2018; pp. 221–279. [CrossRef]

23. Alcaraz, R.; Hornero, F.; Rieta, J.J. Electrocardiographic Spectral Features for Long-Term Outcome Prognosis of Atrial Fibrillation Catheter Ablation. *Ann. Biomed. Eng.* **2016**, *44*, 3307–3318. [CrossRef] [PubMed]

24. Everett, T.H., 4th; Kok, L.C.; Vaughn, R.H.; Moorman, J.R.; Haines, D.E. Frequency domain algorithm for quantifying atrial fibrillation organization to increase defibrillation efficacy. *IEEE Trans. Biomed. Eng.* **2001**, *48*, 969–978. [CrossRef] [PubMed]

25. Holmqvist, F.; Stridh, M.; Waktare, J.E.P.; Roijer, A.; Sörnmo, L.; Platonov, P.G.; Meurling, C.J. Atrial fibrillation signal organization predicts sinus rhythm maintenance in patients undergoing cardioversion of atrial fibrillation. *Europace* **2006**, *8*, 559–565. [CrossRef]

26. Husser, D.; Stridh, M.; Sornmo, L.; Geller, C.; Klein, H.U.; Olsson, S.B.; Bollmann, A. Time-frequency analysis of the surface electrocardiogram for monitoring antiarrhythmic drug effects in atrial fibrillation. *Am. J. Cardiol.* **2005**, *95*, 526–528. [CrossRef] [PubMed]

27. Michaud, G.; Kumar, S. Pulmonary vein isolation in the treatment of atrial fibrillation. *Res. Rep. Clin. Cardiol.* **2016**, *7*, 47–60. [CrossRef]

28. Morin, D.P.; Bernard, M.L.; Madias, C.; Rogers, P.A.; Thihalolipavan, S.; Estes, N.A.M., 3rd. The State of the Art: Atrial Fibrillation Epidemiology, Prevention, and Treatment. *Mayo Clin. Proc.* **2016**, *91*, 1778–1810. [CrossRef]

29. Petrénas, A.; Marozas, V.; Sörnmo, L. Lead Systems and Recording Devices. In *Atrial Fibrillation from an Engineering Perspective*; Springer International Publishing: Berlin, Germany, 2018; pp. 25–48. [CrossRef]

30. Sörnmo, L.; Laguna, P. Chapter 7—ECG Signal Processing. In *Bioelectrical Signal Processing in Cardiac and Neurological Applications*; Sörnmo, L., Laguna, P., Eds.; Biomedical Engineering, Academic Press: Burlington, NJ, USA, 2005; pp. 453–566. [CrossRef]

31. García, M.; Martínez-Iniesta, M.; Ródenas, J.; Rieta, J.J.; Alcaraz, R. A novel wavelet-based filtering strategy to remove powerline interference from electrocardiograms with atrial fibrillation. *Physiol. Meas.* **2018**, *39*, 115006. [CrossRef]

32. Alcaraz, R.; Rieta, J.J. Adaptive singular value cancelation of ventricular activity in single-lead atrial fibrillation electrocardiograms. *Physiol. Meas.* **2008**, *29*, 1351–1369. [CrossRef]

33. Henriksson, M.; García-Alberola, A.; Sörnmo, L. Short-term reproducibility of parameters characterizing atrial fibrillatory waves. *Comput. Biol. Med.* **2020**, *117*, 103613. [CrossRef]

34. Murase, Y.; Inden, Y.; Shibata, R.; Yanagisawa, S.; Fujii, A.; Ando, M.; Otake, N.; Takenaka, M.; Funabiki, J.; Sakamoto, Y.; et al. The impact of the dominant frequency of body surface electrocardiography in patients with persistent atrial fibrillation. *Heart Vessel.* **2020**, *35*, 967–976. [CrossRef]

35. Szilágyi, J.; Walters, T.E.; Marcus, G.M.; Vedantham, V.; Moss, J.D.; Badhwar, N.; Lee, B.; Lee, R.; Tseng, Z.H.; Gerstenfeld, E.P. Surface ECG and intracardiac spectral measures predict atrial fibrillation recurrence after catheter ablation. *J. Cardiovasc. Electrophysiol.* **2018**, *29*, 1371–1378. [CrossRef]

36. Lankveld, T.; Zeemering, S.; Scherr, D.; Kuklik, P.; Hoffmann, B.A.; Willems, S.; Pieske, B.; Haïssaguerre, M.; Jaïs, P.; Crijns, H.J.; et al. Atrial Fibrillation Complexity Parameters Derived From Surface ECGs Predict Procedural Outcome and Long-Term

Follow-Up of Stepwise Catheter Ablation for Atrial Fibrillation. *Circ. Arrhythm. Electrophysiol.* **2016**, *9*, e003354. [CrossRef] [PubMed]

37. Eftestol, T.; Sunde, K.; Ole Aase, S.; Husoy, J.H.; Steen, P.A. Predicting outcome of defibrillation by spectral characterization and nonparametric classification of ventricular fibrillation in patients with out-of-hospital cardiac arrest. *Circulation* **2000**, *102*, 1523–1529. [CrossRef] [PubMed]

38. Löfhede, J.; Thordstein, M.; Löfgren, N.; Flisberg, A.; Rosa-Zurera, M.; Kjellmer, I.; Lindecrantz, K. Automatic classification of background EEG activity in healthy and sick neonates. *J. Neural Eng.* **2010**, *7*, 16007. [CrossRef] [PubMed]

39. Uldry, L.; Van Zaen, J.; Prudat, Y.; Kappenberger, L.; Vesin, J.M. Measures of spatiotemporal organization differentiate persistent from long-standing atrial fibrillation. *Europace* **2012**, *14*, 1125–1131. [CrossRef] [PubMed]

40. Julián, M.; Alcaraz, R.; Rieta, J.J. Comparative assessment of nonlinear metrics to quantify organization-related events in surface electrocardiograms of atrial fibrillation. *Comput. Biol. Med.* **2014**, *48*, 66–76. [CrossRef] [PubMed]

41. Xiong, P.Y.; Jahanshahi, H.; Alcaraz, R.; Chu, Y.M.; Gómez-Aguilar, J.; Alsaadi, F.E. Spectral entropy analysis and synchronization of a multi-stable fractional-order chaotic system using a novel neural network-based chattering-free sliding mode technique. *Chaos Solitons Fractals* **2021**, *144*, 110576. [CrossRef]

42. Shen, E.; Cai, Z.; Gu, F. Mathematical foundation of a new complexity measure. *Appl. Math. Mech.* **2005**, *26*, 1188–1196. [CrossRef]

43. Refaeilzadeh, P.; Tang, L.; Liu, H. Cross-Validation. In *Encyclopedia of Database Systems*; Liu, L., Ozsu, M.T., Eds.; Springer: Berlin, Germany, 2009; pp. 532–538. [CrossRef]

44. Izenman, A.J. Linear Discriminant Analysis. In *Springer Texts in Statistics*; Springer: New York, NY, USA, 2013; pp. 237–280. [CrossRef]

45. Habibzadeh, F.; Habibzadeh, P.; Yadollahie, M. On determining the most appropriate test cut-off value: The case of tests with continuous results. *Biochem. Med.* **2016**, *26*, 297–307. [CrossRef]

46. Rückstieß, T.; Osendorfer, C.; van der Smagt, P. Sequential Feature Selection for Classification. In *AI 2011: Advances in Artificial Intelligence*; Springer: Berlin/Heidelberg, Germany, 2011; pp. 132–141. [CrossRef]

47. Fagerland, M.W.; Lydersen, S.; Laake, P. The McNemar test for binary matched-pairs data: Mid-p and asymptotic are better than exact conditional. *BMC Med. Res. Methodol.* **2013**, *13*, 91. [CrossRef]

48. Gerstenfeld, E.P.; Duggirala, S. Atrial fibrillation ablation: Indications, emerging techniques, and follow-up. *Prog. Cardiovasc. Dis.* **2015**, *58*, 202–212. [CrossRef] [PubMed]

49. Dretzke, J.; Chuchu, N.; Agarwal, R.; Herd, C.; Chua, W.; Fabritz, L.; Bayliss, S.; Kotecha, D.; Deeks, J.J.; Kirchhof, P.; et al. Predicting recurrent atrial fibrillation after catheter ablation: A systematic review of prognostic models. *Europace* **2020**, *22*, 748–760. [CrossRef] [PubMed]

50. Lambert, L.; Marek, J.; Fingrova, Z.; Havranek, S.; Kuchynka, P.; Cerny, V.; Simek, J.; Burgetova, A. The predictive value of cardiac morphology for long-term outcome of patients undergoing catheter ablation for atrial fibrillation. *J. Cardiovasc. Comput. Tomogr.* **2018**, *12*, 418–424. [CrossRef]

51. Haïssaguerre, M.; Sanders, P.; Hocini, M.; Hsu, L.F.; Shah, D.C.; Scavée, C.; Takahashi, Y.; Rotter, M.; Pasquié, J.L.; Garrigue, S.; et al. Changes in atrial fibrillation cycle length and inducibility during catheter ablation and their relation to outcome. *Circulation* **2004**, *109*, 3007–3013. [CrossRef] [PubMed]

52. Di Marco, L.Y.; Raine, D.; Bourke, J.P.; Langley, P. Atrial Fibrillation Type Characterization and Catheter Ablation Acute Outcome Prediction: Comparative Analysis of Spectral and Nonlinear Indices from Right Atrium Electrograms. In Proceedings of the 41st Computing in Cardiology Conference (CinC), Cambridge, MA, USA, 7–10 September 2014; Volume 41, pp. 817–820.

53. Borowska, M. Entropy-based algorithms in the analysis of biomedical signals. *Stud. Logic Gramm. Rhetor.* **2015**, *43*, 21–32. [CrossRef]

54. Sanchez-Somonte, P.; Jiang, C.Y.; Betts, T.R.; Chen, J.; Mantovan, R.; Macle, L.; Morillo, C.A.; Haverkamp, W.; Weerasooriya, R.; Albenque, J.P.; et al. Completeness of Linear or Fractionated Electrogram Ablation in Addition to Pulmonary Vein Isolation on Ablation Outcome: A Substudy of the STAR AF II Trial. *Circ. Arrhythm. Electrophysiol.* **2021**, *14*, e010146. [CrossRef]

55. Verma, A.; Jiang, C.y.; Betts, T.R.; Chen, J.; Deisenhofer, I.; Mantovan, R.; Macle, L.; Morillo, C.A.; Haverkamp, W.; Weerasooriya, R.; et al. Approaches to catheter ablation for persistent atrial fibrillation. *N. Engl. J. Med.* **2015**, *372*, 1812–1822. [CrossRef]

56. Cui, X.; Chang, H.C.; Lin, L.Y.; Yu, C.C.; Hsieh, W.H.; Li, W.; Peng, C.K.; Lin, J.L.; Lo, M.T. Prediction of atrial fibrillation recurrence before catheter ablation using an adaptive nonlinear and non-stationary surface ECG analysis. *Phys. A Stat. Mech. Its Appl.* **2019**, *514*, 9–19. [CrossRef]

57. Hsu, N.W.; Lin, Y.J.; Tai, C.T.; Kao, T.; Chang, S.L.; Wongcharoen, W.; Lo, L.W.; Udyavar, A.R.; Hu, Y.F.; Tso, H.W.; et al. Frequency analysis of the fibrillatory activity from surface ECG lead V1 and intracardiac recordings: Implications for mapping of AF. *Europace* **2008**, *10*, 438–443. [CrossRef]

58. Hidalgo-Muñoz, A.R.; Latcu, D.G.; Meo, M.; Meste, O.; Popescu, I.; Saoudi, N.; Zarzoso, V. Spectral and spatiotemporal variability ECG parameters linked to catheter ablation outcome in persistent atrial fibrillation. *Comput. Biol. Med.* **2017**, *88*, 126–131. [CrossRef]

Journal of
*Personalized
Medicine*

Article

Comparative Study of Methods for Cycle Length Estimation in Fractionated Electrograms of Atrial Fibrillation

Diego Osorio [1], Aikaterini Vraka [1], José Moreno-Arribas [2], Vicente Bertomeu-González [2], Raúl Alcaraz [3] and José J. Rieta [1,*]

[1] BioMIT.org, Electronic Engineering Department, Universitat Politecnica de Valencia, 46022 Valencia, Spain
[2] Cardiology Department, Saint John's University Hospital, 03550 Alicante, Spain
[3] Research Group in Electronic, Biomedical and Telecommunication Engineering, University of Castilla-La Mancha, 16071 Cuenca, Spain
* Correspondence: jjrieta@upv.es

Abstract: Atrial cycle length (CL) is an important feature for the analysis of electrogram (EGM) characteristics acquired during catheter ablation (CA) of atrial fibrillation (AF), the commonest cardiac arrhythmia. Nevertheless, a robust ACL estimator requires the precise detection of local activation waves (LAWs), which still remains a challenge. This work aims to compare the performance in (CL) estimation, especially under fractionated EGMs, of three different LAW detection methods relying on different operation strategies. The methods are based on the hyperbolic tangent (HT) function, an adaptive amplitude threshold (AAT) and a (CL) iteration (ACLI), respectively. For each method, LAW detection has been assessed with respect to manual annotations made by two experts and performance has been estimated by confusion matrix and mean and individual (CL) error calculation by EGM types of fractionation. The influence of EGM length on the individual (CL) error has been additionally considered. For the HT method, accuracy, sensitivity and precision were 92.77–100%, while for the AAT and ACLI methods they were 78.89–99.91% for all EGM types. The CL error on the HT method was lower than AAT and ACLI methods (up to 12 ms versus up to 20 ms), with the difference being more prominent in complex EGMs. The HT method also showed the lowest dependency on EGM length, presenting the lowest and least variable error values. Therefore, the HT method achieves higher performance in (CL) estimation in comparison with previous LAW detection techniques. The high robustness and precision demonstrated by this method suggest its implementation on CA mapping devices for a more successful location of ablation targets and improved results during CA procedures.

Keywords: atrial fibrillation; electrogram; complex fractionated atrial electrograms; local activation waves; detection; invasive recordings; cycle length; comparison

Citation: Osorio, D.; Vraka, A.; Moreno-Arribas, J.; Bertomeu-González, V.; Alcaraz, R.; Rieta, J.J. Comparative Study of Methods for Cycle Length Estimation in Fractionated Electrograms of Atrial Fibrillation. *J. Pers. Med.* **2022**, *12*, 1712. https://doi.org/10.3390/jpm12101712

Academic Editor: José Miguel Rivera-Caravaca

Received: 25 August 2022
Accepted: 11 October 2022
Published: 13 October 2022

1. Introduction

Being the most common cardiac arrhythmia in clinical practice, atrial fibrillation (AF) affects more than 43 million people worldwide [1]. The risks of stroke, myocardial infarction, heart failure or dementia are increased in patients suffering from AF, which is also associated with increased mortality [1,2]. The ever-growing life expectancy in comparison to the high AF prevalence for older individuals [1] and the significant economic burden that AF provokes [1,3] highlight the need for an efficient treatment.

So far, catheter ablation (CA) is considered the star AF treatment due to its high effectiveness and safety [1,4,5]. The aim of the CA procedure is to eliminate the ectopic atrial beats that can trigger or sustain AF by electrically isolating the areas that are thought to present arrhythmogenic activity or fibrosis, respectively [1,6]. Pulmonary veins (PVs) are the most common AF foci, especially for patients in paroxysmal AF [1,7,8]. Nevertheless, other atrial sites may contribute to AF perpetuation, especially in persistent AF patients, where the atrial remodeling takes place to a higher extent [1,9]. Consequently, additional CA may be necessary in order to achieve freedom from AF [10]. Areas targeted for additional CA are defined via electrogram (EGM) analysis, which allows the detection of remodeled tissue favoring AF perpetuation or even triggering AF. Low-voltage, high dominant frequency (DF), low AF cycle length (CL) or highly complex EGMs, called complex fractionated atrial electrograms (CFAEs), are some of the most established techniques to define candidate CA targets, albeit showing controversial results [11–16]. These types of EGMs are thought to correspond to areas of slow conduction or pivotal points of reentrant wavelets, revealing the presence of atrial fibrosis which provokes AF perpetuation [17].

For many of the aforementioned techniques, the detection of local activation waves (LAWs) is an elemental step. In low fractionated EGMs, LAWs can be easily detected, hence facilitating the analysis [18]. However, in CFAEs and highly fractionated EGMs, LAWs detection can become a hard task often misleading the CL estimation and, consequently, the procedures and assessments based on this feature [13]. Moreover, the possibility of ambiguous LAWs annotations in CFAEs may be an explanation for the low impact of non-PV CA on the freedom from AF [14,15]. Therefore, the development and disposal of a reliable and robust LAW estimator are of paramount importance. Although various approaches to the development of LAWs detectors have been attempted so far [19–27], most of them focus on unipolar EGMs [20,21,25–27], where LAWs detection is a much easier task [13]. Nevertheless, as LAWs detection in bipolar EGMs is more complicated due to the nonstationary nature of the recorded dynamics as well as the high dependency on the wavefront direction, not but a few techniques for bipolar EGMs have been developed so far [19,22–24].

One of the first and most cited works employed an adaptive amplitude-based threshold which varied according to the last detected peaks, showing low error values for regular or low fractionated EGMs but increased rates for fractionated signals [19]. The CL-based method is another highlighted work [22]. Search is performed with an initial amplitude threshold which is then modified according to CL. This method has also shown low error rates for regular EGMs. Nevertheless, when CL becomes irregular, as normally happens in CFAEs, error rates increase. The DF-based method detects LAWs corresponding to periodic peaks of the EGMs [23]. Although this method showed good performance compared to previous methods regarding periodic EGMs, it also failed to show satisfactory results for aperiodic signals, which is a much more frequent event in highly fractionated EGMs. Thus, it will not be considered in this study. Finally, the newer of the aforementioned methods is based on the application of the hyperbolic tangent (HT) function on CFAEs in order to highlight and facilitate low amplitude detections as well as to limit very high amplitude activations using a hybrid methodology combining attributes of the CL- and adaptive amplitude-methods, showing high performance in CFAEs and complex EGMs [24].

Given the advantage of bipolar over unipolar EGMs in being insensitive to ventricular activity and the high use of the former in AF studies, the appropriate strategy choice for works that may need to implement LAWs detection in their analysis can be quite confusing yet inevitable. Despite the results shown in the aforementioned studies, each one of them has used a different database. Hence, a direct comparison would be unfair or would not be directly performed. The main objective of the present work is to perform a straightforward comparison between the three most robust aforementioned LAW detectors, the adaptive amplitude threshold (AAT) [19], the atrial CL iterator (ACLI) method [22] and the hyperbolic tangent (HT) method [24], by reproducing each one of them and comparing the results over the same database. Since the common denominator of LAWs detection in any of the aforementioned studies is the CL estimation, using this parameter as a performance evaluation metric seems a fair option. Comparison is performed over a range of factors that could affect or create biased results, such as the length of EGMs or different fractionation levels. In this way, a multidimensional assessment can be performed in order to offer a more reliable and direct comparative measure for future studies.

The manuscript is organized as follows. Section 2 presents the dataset used in this study and the preprocessing methods applied, as well as a brief presentation of the methodology of the strategies under comparison and the evaluation methods used. Section 3 includes the results of the comparison, while Section 4 provides an overview of the methods under comparison and analyzes the main conclusions derived from the results, which are then summarized in Section 5.

2. Methods

The dataset employed in this study consisted of 119 EGMs of 10 s in length, recorded by a Labsystem™ PRO EP recording system (Boston Scientific, Marlborough, MA, USA) after obtaining the written consent of 22 patients undergoing CA procedures. EGMs were sampled at 1 kHz and filtered by an adaptive notch filter to reduce powerline interference together with a 0.5–500 Hz band-pass filter to limit high-frequency noise. The dataset was manually annotated and classified by EGM types by two expert physicians according to Wells' classification [28]. Type I EGMs consisted of clear LAWs and almost isoelectric baseline, type II contained low perturbations in the baseline and a higher number of deflections at each LAW while type III contained a high fluctuating baseline and low-amplitude fragmented LAWs. Sixteen EGMs were classified by the experts as AF type I, 19 as AF type II and 84 as AF type III, so that the vast majority of the EGMs selected pertained to the most difficult AF type to delineate. Signals were denoised by a Wavelet Transform-based technique, which has been proved to perform better than regular filtering, providing quick results, effective noise elimination and optimal preservation of the signal morphology [29].

2.1. LAW Detection Methods

The methodology of each of the studied LAW detection strategies is presented briefly in the next subsections, followed by a comparative example of the preprocessing method adopted by each strategy.

2.1.1. Adaptive Amplitude Threshold Method

The first method implemented was the AAT algorithm [19]. This technique is based on Botteron and Smith preprocessing [18], a technique developed to generate waveforms proportional to the amplitude of the EGM components with frequencies within the band-pass filter cutoffs. This method includes 40–250 Hz band-pass filtering, followed by signal rectification and low-pass filtering with the cut-off frequency at 20 Hz. The AAT algorithm employs an AAT based on the last ten detected activations with exponentially decreasing weights (s_w), resulting from the application of the Botteron filtering, using a blanking period of 55 ms between activations. Lastly, the modulus of the original signal is filtered by a non-causal moving average filter with 90 coefficients, resulting in a new signal s_f, where the positive zero crossings closer to a local peak of s_w define the activation times. A more detailed description of the algorithm can be found elsewhere [19].

2.1.2. Atrial Cycle Length Iteration Method

The ACLI method [22] is based on a modified Botteron and Smith [18] preprocessing, replacing the band-pass filter with a 40 Hz high-pass filter and increasing the low-pass filter cut-off to 30 Hz. The algorithm performs an iterative LAW detection, starting with the highest amplitude peak and moving to the next peak according to an amplitude descending order, making use of a blanking period of 50 ms, until the average CL is lower than 275 ms. Once this condition is fulfilled, a second condition has to be accomplished so that the iteration stops: either the mean CL has to be less than the median CL plus 5 ms or the amplitude of the current peak has to be 20% less than the amplitude of the previous peak. Finally, a loop control checks the existence of intervals longer than 1.5 × the median CL and adds, in case such intervals are found, the highest peaks within these intervals to the activation set. More details on the methodology can be found elsewhere [22].

2.1.3. Hyperbolic Tangent Method

The HT method [24] also modifies the Botteron and Smith preprocessing steps, expanding the range of the bandpass filter cutoffs from 40–250 to 20–250 Hz. Studies on the original preprocessing effect on CFAEs demonstrated that slower components of activations were being lost, provoking an insufficient waveform response, leading the algorithm to miss the respective activations [24]. Therefore, a reduction in the minimum band-pass filter cut-off was necessary, with 20 Hz being the optimal frequency based on analyses performed.

The preprocessing of this method also used a novel technique, the HT function, aiming to reduce high amplitude variability observed in CFAEs in order to facilitate the detection process that follows next. For this purpose, each recording is segmented into 7-second intervals with 25% overlapping and classified according to their fractionation level by their kurtosis calculated at each 1 s and averaged, so that the HT function is applied exclusively on CFAEs. The detection process is initiated by an amplitude-based localization of all peaks higher than 0.4 mV, taking into consideration the minimum refractory period of 50 ms. The process continues with the median CL calculation and a secondary search on intervals longer than the median CL is performed, reducing the amplitude threshold proportionally to the difference between the median CL and the current interval. The process finishes with the calculation of the barycenter of each LAW, defined as the sample point that equally divides the area of the modulus of the LAWs, in order to estimate the activation time. More details on this algorithm are provided elsewhere [24].

A comparative example of these three different preprocessing strategies applied over the same EGM is shown in Figure 1, where it can be observed how the HT method produces a signal in which activations are easier to detect automatically in later stages because small peaks are enhanced and large peaks are limited. Additionally, block diagrams of each technique's methodology are presented in Figure 2.

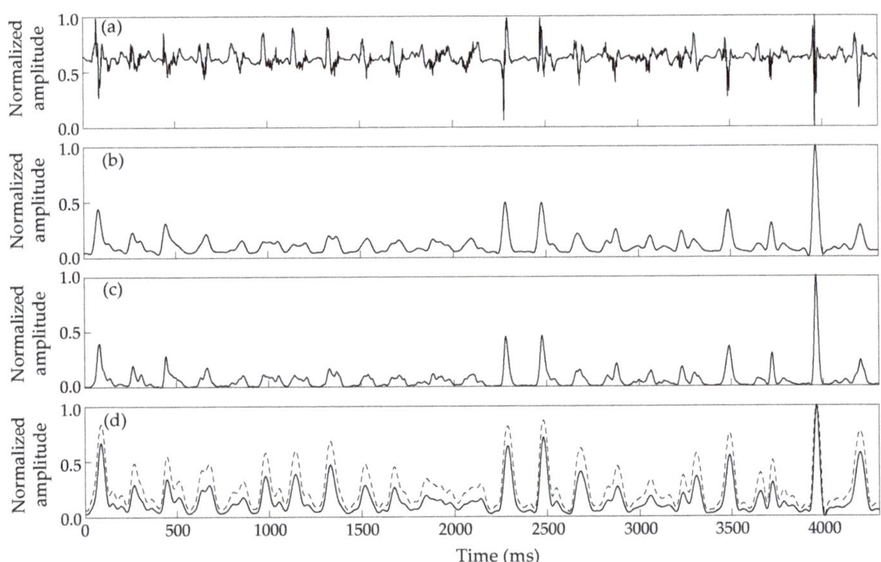

Figure 1. Comparative analysis of the preprocessing stage of each of the compared LAW detection methods applied on the same EGM: (**a**) Denoised EGM to be processed; (**b**) Result of AAT preprocessing; (**c**) Result of ACLI preprocessing; (**d**) Result of HT preprocessing. The dashed line in (**d**) corresponds to the application of the hyperbolic tangent.

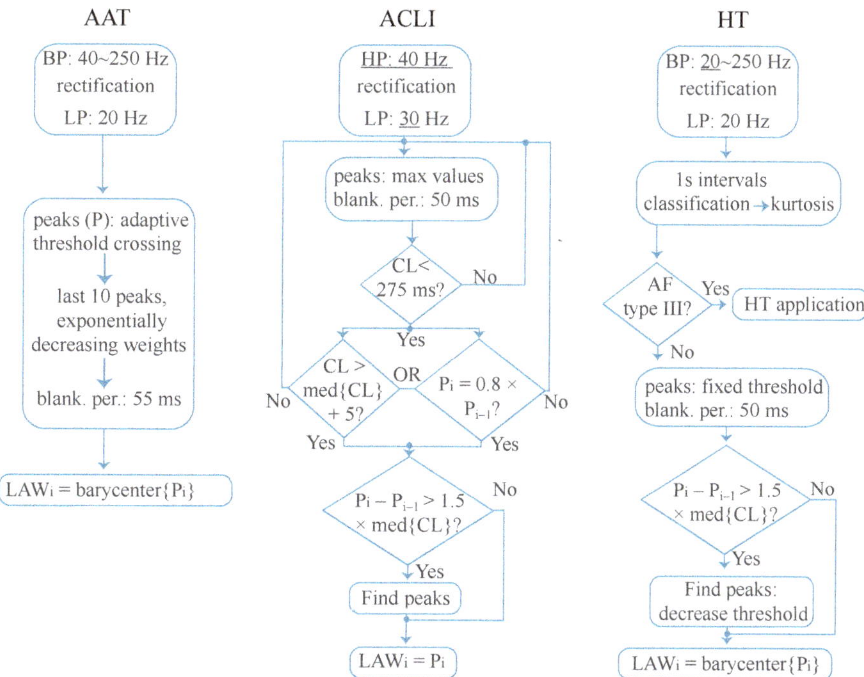

Figure 2. Block diagrams of the LAW detection methods compared. Underlined values in the first blocks show modified thresholds with respect to traditional Botteron's preprocessing method. BP: band-pass filtering; LP: low-pass filtering; HP: high-pass filtering; P_i: amplitude of the *i*-th peak; med{CL}: median cycle length.

2.2. Performance Evaluation

Confusion matrix and CL error metrics were employed in order to evaluate the performance. Regarding the confusion matrix, *Accuracy*, *Sensitivity* and *Precision* were calculated. True positive (TP) were the LAWs annotated both by the experts and by the method under analysis. False positives (FP) were the LAWs that were only annotated by the method, while false negatives (FN) were the LAWs that were annotated only by the experts. Since there are no true negatives (TN), these metrics were calculated as follows:

$$Accuracy = \frac{TP}{TP + FP + FN},\tag{1}$$

$$Sensitivity = \frac{TP}{TP + FN},\tag{2}$$

and

$$Precision = \frac{TP}{TP + FP}.\tag{3}$$

Details on CL error calculation are provided next. The effect of EGMs duration as well as fractionation has also been considered.

2.2.1. Detection Performance

The number of LAWs successfully detected at each method was calculated by comparing the annotations made by the algorithm with the manual annotations made by the experts, allowing a threshold of 40 ms. Performance was evaluated in terms of Accuracy, Sensitivity and Precision, providing an overall perspective on performance, including the undersensing and oversensing rates of each algorithm.

2.2.2. Mean CL Error Estimation

The error in estimating the average CL at any of the three methods was measured as the absolute difference between the mean CL of the manual and algorithm's annotations, as can be seen from Figure 3, which will be explained more in-depth later for individual CL error estimation. Nevertheless, considering that oversensing or undersensing can have an effect on the mean CL, producing misleading results as can be seen in Figure 3B, mean CL error can not be considered as a reliable performance indicator. Thus, an analysis of individual CL errors was additionally developed in order to obtain more realistic results.

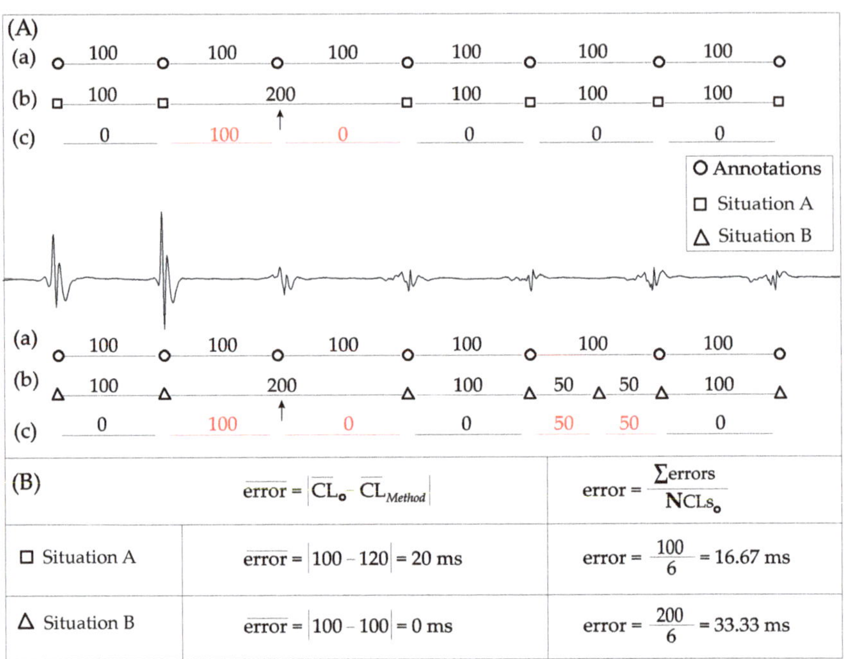

Figure 3. Comparative example for calculating the CL estimation error. (**A**) EGM with experts' (circle) and method's annotations: (a) expert's CL (b) method's CL and (c) errors in CL estimation. Top: Situation 1 (squares) undersensing case (missing activation-red). The arrow shows the starting points in case of unequal CLs for the undersensing case; Bottom: Situation 2 (triangles) undersensing and oversensing case (false positive activation-red). (**B**) Error estimation for the two example situations. (left) Mean CL error; (right) Individual CL error. Notice how the simultaneous existence of under- and oversensing in Situation 2 may create erroneous CL calculation for the mean CL case.

2.2.3. Individual CL Error Estimation

As previously stated, this metric was introduced in order to reliably calculate the CL error for each algorithm, without any bias inserted from possible under- or oversensing tendencies. In this case, the error at each CL is defined by the absolute difference between the manual annotations, called reference annotations, and the algorithm's annotations. The individual CL error is finally calculated by the sum of errors found at each CL divided by the number of reference intervals. The formula for the calculation of the individual CL error is given in the right column of Figure 3B. If the CL was the same for the manual and algorithm annotations, the respective error was set to zero. However, if there was some under- or oversensing detected, the evaluation process was able to quantify the error.

For the undersensing case, as the given CL was longer for the algorithm than for the manual annotations case, the corresponding error is calculated by subtracting the reference CL from the algorithm's CL. However, in that case, the next interval under comparison will involve the following reference segment, while the algorithm's segment will remain the same as before, considered to be starting from the reference segment's initiation point so that a double counting on error will be avoided. An example of this process can be seen in Figure 3A, with experts' (circle) and method's (square, triangle) annotations. The top panel shows Situation 1 (square) with an undersensing case where the arrow shows the beginning of the algorithm's segment (missed activation-red). For each interval between annotations, line (a) indicates the expert's CL, line (b) the method's CL and line (c) errors in the estimation of each CL. The bottom panel shows Situation 2 (triangles) with both undersensing (missed activation-red) and oversensing cases (falsely detected activation-red). Regarding the oversensing case, the error at the given interval will be calculated by subtracting the algorithm's CL from the reference CL, for each one of the algorithm's intervals between a reference interval, as can be seen from Situation 2 of Figure 3A. Finally, as the mean CL of Situation 1, with one missed activation, is 120 ms, then the mean error is 20 ms. Nevertheless, the simultaneous existence of under- and oversensing in Situation 2 will create an erroneous mean CL calculation of 0 ms, as this method had a mean CL of 100 ms, even though not all the activations corresponded to the experts' annotations. On the contrary, the individual error of Situation 2 was 33.33 ms, both indicating a non-perfect annotator and showing a higher error than Situation 1 (16.67 ms), which only had one mishit.

2.2.4. Influence of Electrograms Duration

Another aspect analyzed in this study was the influence of the EGMs duration on the individual CL error of each method. For each type of EGM, the three methods were applied progressively from 1 to 10 s of duration over the dataset, computing the individual CL error as a measure of the CL estimation precision.

3. Results

3.1. Detection Results

An example of the three AF-type EGMs along with LAWs annotations from experts and the compared techniques can be seen in Figure 4. While LAWs detection is pretty straightforward for AF types I and II, LAWs detection in AF type III EGMs is significantly complicated, with each algorithm presenting some false positive or negative detections. Results on detection performance for the methods under comparison are shown in Table 1 for EGMs of all AF types. Both ACLI and HT methods show optimal results for types I and II. On the contrary, the AAT method showed lower accuracy and precision rates, which is especially unusual for AF type I. This fact is provoked by the high variability in amplitude observed in several type I EGMs of the dataset used for the present analysis. Figure 3 corresponds to a type I EGM example, in which the impact of amplitude variability on the AAT method's detection performance can be seen. Although AAT's accuracy rate is not lower than 90% in type I EGMs, this relatively low performance implies significant deficiencies of the AAT method and a higher CL error, provoked by the longer CL of AF type I EGMs, leading to the lower number of intervals and, consequently, the higher weight of each mishit.

Table 1. Electrogram detection performance results for all AF types of the three LAW detectors compared in this study in terms of Accuracy (*Acc*), Sensitivity (*Se*) and Precision (*Pr*).

Method	Type I			Type II			Type III		
	Acc [%]	*Se* [%]	*Pr* [%]	*Acc* [%]	*Se* [%]	*Pr* [%]	*Acc* [%]	*Se* [%]	*Pr* [%]
AAT	91.82	95.60	95.66	95.90	97.35	98.33	78.89	85.38	91.11
ACLI	98.58	99.91	98.68	98.78	99.73	99.05	85.13	87.51	96.89
HT	100.00	100.00	100.00	100.00	100.00	100.00	92.77	95.30	97.24

The results for type III EGMs, also shown in Table 1, are of great importance in understanding the performance of each method, as this type of EGM coincides with the most fragmented signals and CFAEs. Firstly, the AAT method shows the lowest accuracy results, while sensitivity and precision are quite compensated, with values of 85.38% and 91.11%, respectively, due to higher undersensing than oversensing incidents. The ACLI method shows improved results in comparison with the AAT method. However, a quite intense discrepancy can be seen between undersensing and oversensing tendencies of this method, as can be observed from the difference between sensitivity and precision rates at 87.51% and 96.89%, respectively. Finally, the HT method presents the most satisfactory results for all confusion matrix parameters, with a balance observed between undersensing and oversensing tendencies, expressed by the 95.30% and 97.24% of sensitivity and precision values, respectively, while showing the highest accuracy rates as well.

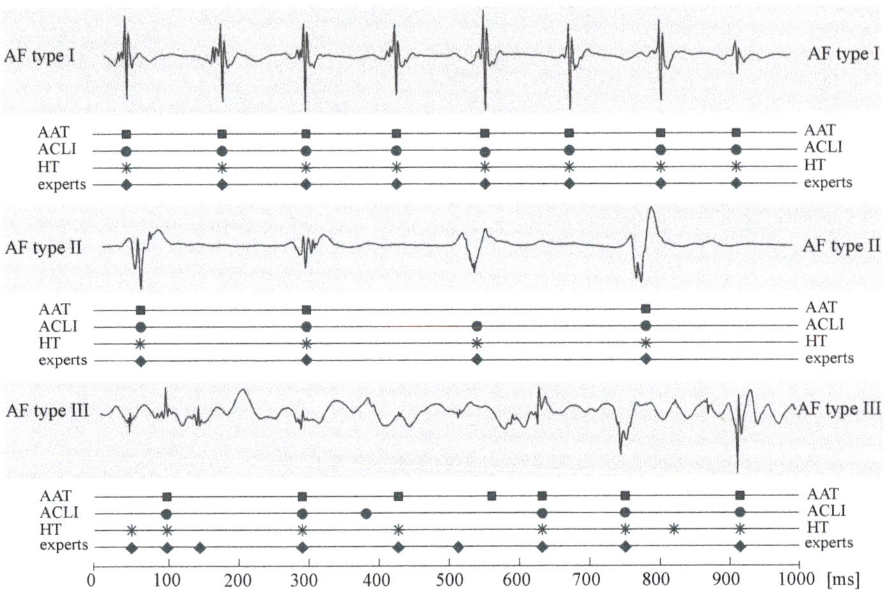

Figure 4. One-second example of the three EGM types and the annotations made from the three algorithms as well as from the experts. Experts' annotations are considered as ground truth. For AF type I, activations were clearly distinguished from the baseline and all the methods were able to identify them. For AF type II, although the baseline shows little perturbation, lower amplitude activations slightly complicate the procedure, leading to a mishit for the AAT algorithm. Finally, for AF type III EGMs, mishits and false positives are observed for all methods across the EGM.

3.2. Cycle Length Estimation

Regarding CL results, Figure 5a shows the mean and individual CL errors on type I EGMs. The mean CL error in ACLI and HT methods is close to 0, as expected from the high detection accuracy and the equilibrium between sensitivity and precision. Individual CL error in these methods is also quite low, yet slightly higher than mean CL error since bias from undersensing or oversensing behavior is removed in this parameter. Regarding AAT results, both mean and individual CL error are significantly higher, at about 10 ms and 12 ms, respectively. This is explained by the relatively low accuracy of this method in type I EGMs. As before, higher individual CL errors are explained by the balance between the undersensing and oversensing rates of these EGM types.

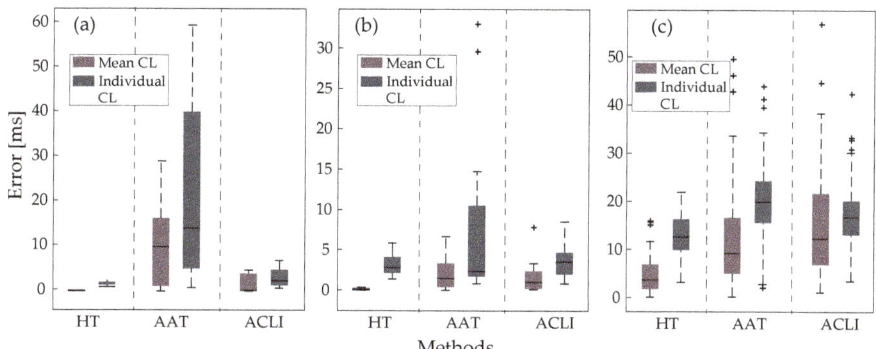

Figure 5. Comparative analysis of errors estimating CL using the mean CL and the individual CL methods for the three LAW detectors compared: (**a**) Results for type I electrograms; (**b**) type II electrograms; and (**c**) type III electrograms. Values indicated with + stand for outliers.

The results for type II EGMs are presented in Figure 5b. ACLI and HT methods show results similar to those from type I EGMs, with an increased difference between individual and mean CL errors. As previously mentioned, mean CL estimation for the AAT method compensated for the errors provoked by under- or oversensing. Lower amplitude variability in the employed dataset led to better performance in AF type II than type I EGMs for the AAT method, as was observed from the detection results. According to this performance improvement, the error in type II EGMs was lower than type I for the AAT method.

Finally, Figure 5c shows the CL error on type III EGMs, which include CFAEs. Being the most complicated case, analysis of this type of EGM clearly demonstrates that individual CL error estimation is a more suitable parameter, efficiently removing any impact of external factors on the CL error calculation, as can be seen from the discrepancies between the mean and individual CL error estimation of each method, leading to invalid conclusions in the case of the AAT and ACLI methods. Error on all methods is higher in this EGM type, with the HT technique showing the best results both in mean (about 4 ms in HT vs. 9 ms in AAT vs. 11 ms in ACLI) and individual (about 12 ms in HT vs. 20 ms in AAT vs. 17 ms in ACLI) CL error analysis.

3.3. EGM Length Influence

The results of the application of each method over the dataset for different EGM lengths are presented in Figure 6, where individual CL error is analyzed. The HT method shows the most stable results regarding length, being additionally improved in comparison with the other methods. For type I EGMs, the ACLI method shows similar results. As the fractionation level increases, however, ACLI performance falls, showing significantly worse results. Finally, AAT error is the highest regarding all AF types, while this method also shows a higher dependency on the EGM length.

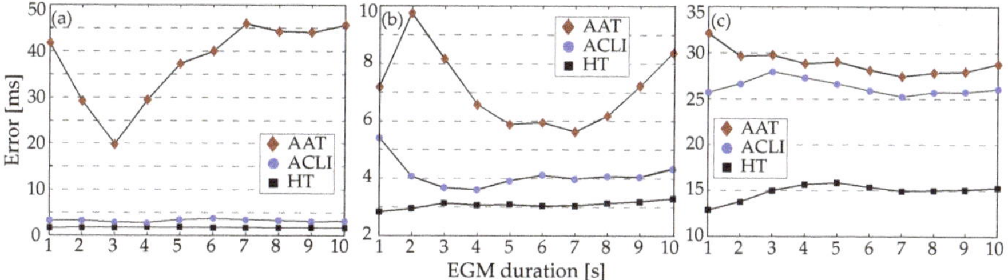

Figure 6. Individual CL error depending on EGM length for the three LAW detection methods analyzed: (**a**) Results for type I electrograms; (**b**) type II electrograms; and (**c**) type III electrograms.

4. Discussion

Due to the high importance of precise AF substrate mapping, several methods for LAW detection have been developed [19,20,22–26] until the present. However, some of these techniques cannot be reproduced or directly applied on CA devices, due to either the particularity of the database [20] used or to the controversial methodology followed, using computational models premised on assumptions about tissue's geometry [25,26]. Other techniques can be directly implemented on substrate mapping devices, such as the AAT [19], the ACLI [22], the dominant frequency (DF) [23] and the HT [24] methods. Each one of these studies has shown promising results, the correct choice for future works in need of a reliable LAWs detector, though, can be confusing as the database used is different.

The present study has performed a direct comparison of three of the aforementioned methods, the AAT, the ACLI and the HT method. The reason why the DF method was not included in the comparison is the lack of individual CL estimation of the DF method, which is one of the main evaluation parameters of the present study. In addition, the DF method failed to provide satisfactory results under aperiodic signals, which is the most typical situation in highly fractionated EGMs.The results indicate that HT is the most robust and reliable method, showing higher accuracy and precision and lower CL errors, suggesting its recruitment for both future studies and CA devices.

The high performance of the HT method can be assigned to the alternative preprocessing strategy and the recruitment of the HT function, which prevented the detector from missing low amplitude LAWs in CFAEs. What is more, this method also adopted a segmentation strategy that allowed the adaptation to various signal lengths, while the combination of an adaptive amplitude threshold with CL correction allowed the detection of LAWs in segments with high amplitude variability [24]. Finally, activation time estimation through the barycenters led to an estimation closer to the experts' annotation.

4.1. Evaluation on Different Levels of Fractionation

EGM fractionation plays a significant role in the performance of each method. Normally, in low fractionation environments, performance is high and only a few errors are observed. Thus, low performance in low fractionation EGMs implies a lack of reliability. As EGMs become more fractionated, with indiscernible activations, the LAW detector's accuracy tends to drop. However, since CFAEs, which are highly fractionated EGMs, are the main challenge of any LAW detector, the performance of each method on that kind of recording is crucial and serves as an index to evaluate reliability and robustness.

This study has utilized the Wells' classification for types of EGMs, with ascending levels of fractionation from type I to type III EGMs. In this way, insufficient results in type I EGMs indicate a less reliable method, while competent results on type III EGMs affirm the reliability of another method. The final assessment is a trade-off among the performance of each EGM type, with special weight to type III EGMs, which are the distinguishing parameter.

The HT method showed the highest results in terms of a confusion matrix for all EGM types, being considered this way the most reliable method, achieving high performance. On the contrary, the AAT method showed relatively low results in AF type I EGMs, due to the high amplitude variability present in this EGM type of the current database.

4.2. Cycle Length as an Evaluation Parameter

The main implementation of LAWs detection in real-time mapping devices is related to CL estimation. Consequently, the error in CL estimation was included in the evaluation procedure. As it was shown, mean CL error can be affected by under- or oversensing incidence, and hence, its employment is not recommended. Contrarily, individual CL estimation is an impartial method. Individual CL error was the lowest in the HT method for all EGMs, implying again the high reliability and performance of this technique, while the AAT and ACLI methods have shown poor results, especially for type III EGMs.

4.3. Effect of Signal Length

Adaptability and proper function regardless of external factors are keys to optimized performance. As signal length varies from study to study, the ability of an algorithm to perform optimally under any length should be taken into consideration for the evaluation of its performance. Additionally, signal length could affect the conclusions of the conducted comparison. For the aforementioned reasons, the individual CL error on each method was assessed over various EGM lengths.

The robustness of the HT algorithm was proved not only by the fact that the error was minimal for any EGM type and length but also because it was not significantly affected by the signal length. This means that this method can perform regardless of the duration of the database used, providing flexibility both to prospective studies and to CA devices. On the other side, the AAT method showed the highest individual error for all EGM types, especially high for EGM type I. Moreover, this algorithm showed unusually high dependence on EGM length for types I and II EGMs, possibly due to high discrepancies between amplitude values in various segments of the same recording.

5. Conclusions

The HT method is an optimal LAW detector, showing higher performance than the AAT and ACLI techniques in every aspect under investigation. The high accuracy and robustness of the HT method contribute to precise CL length estimation, regardless of the signal length or the EGM fractionation. The high adaptability of this method suggests its implementation on real-time mapping devices aiming to localize CA targets or its adoption by future studies, providing reliable LAW detection results and precise CL estimation.

Author Contributions: Conceptualization, D.O., A.V., R.A. and J.J.R.; methodology, D.O., A.V., R.A. and J.J.R.; software, D.O.; validation, D.O., J.M.-A., V.B.-G., R.A. and J.J.R.; resources, J.M.-A., V.B.-G. and J.J.R.; data curation, D.O., J.M.-A. and V.B.-G.; original draft preparation, D.O. and A.V.; review and editing, D.O., A.V., J.M.-A., V.B.-G., R.A. and J.J.R. All authors have read and agreed to the published version of the manuscript.

Funding: This research has received financial support from public grants PID2021-123804OB-I00 and PID2021-00X128525-IV0 of the Spanish Government with DOI 10.13039/501100011033 jointly with the European Regional Development Fund (EU), and regional grants SBPLY/17/180501/000411 from Junta de Comunidades de Castilla-La Mancha and AICO/2021/286 from Generalitat Valenciana.

Institutional Review Board Statement: The study was conducted according to the guidelines of the Declaration of Helsinki, complying with Law 14/2007, of July 3rd, on Biomedical Research and other Spanish regulations and was approved by the Ethical Review Board of the University Hospital of San Juan (San Juan de Alicante, Alicante, Spain) with protocol code 21/046.

Informed Consent Statement: Written informed consent was granted from all the subjects participating in the present research. All acquired data were anonymized before processing.

Data Availability Statement: The data supporting reported results and presented in this study are available on request from the corresponding author.

References

1. Hindricks, G.; Potpara, T.; Dagres, N.; Arbelo, E.; Bax, J.J.; Blomström-Lundqvist, C.; Boriani, G.; Castella, M.; Dan, G.-A.; Dilaveriset, P.E.; et al. 2020 ESC Guidelines for the diagnosis and management of atrial fibrillation developed in collaboration with the European Association for Cardio-Thoracic Surgery (EACTS). *Eur. Heart J.* **2021**, *42*, 373–498. [CrossRef] [PubMed]
2. Zuin, M.; Roncon, L.; Passaro, A.; Bosi, C.; Cervellati, C.; Zuliani, G. Risk of dementia in patients with atrial fibrillation: Short versus long follow-up. A systematic review and meta-analysis. *Int. J. Geriatr. Psychiatry* **2021**, *36*, 1488–1500. [CrossRef] [PubMed]

3. Pathak, R.; Evans, M.; Middeldorp, M.; Mahajan, R.; Mehta, A.; Meredith, M.; Twomey, D.; Wong, C.; Hendriks, J.; Abhayaratna, W.; et al. Cost-Effectiveness and Clinical Effectiveness of the Risk Factor Management Clinic in Atrial Fibrillation: The CENT Study. *JACC Clin. Electrophysiol.* **2017**, *3*, 436–447. [CrossRef] [PubMed]

4. Oral, H.; Knight, B.P.; Tada, H. Pulmonary vein isolation for paroxysmal and persistent atrial fibrillation. *Circulation* **2002**, *11*, 83. [CrossRef]

5. Thrall, G.; Lane, D.; Carroll, D.; Lip, G. Quality of Life in Patients with Atrial Fibrillation: A Systematic Review. *Am. J. Med.* **2006**, *119*, 448.e1–448.e19. [CrossRef] [PubMed]

6. Ioannidis, P.; Zografos, T.; Christoforatou, E.; Kouvelas, K.; Tsoumeleas, A.; Vassilopoulos, C. The Electrophysiology of Atrial Fibrillation: From Basic Mechanisms to Catheter Ablation. *Cardiol. Res. Pract.* **2021**, *2021*, 4109269. [CrossRef]

7. Haïssaguerre, M.; Jaïs, P.; Shah, D.C.; Takahashi, A.; Hocini, M.; Quiniou, G.; Garrigue, S.; Le Mouroux, A.; Le Métayer, P.; Clémenty, J. Spontaneous initiation of atrial fibrillation by ectopic beats originating in the pulmonary veins. *N. Engl. J. Med.* **1998**, *339*, 659–666. [CrossRef] [PubMed]

8. Lau, D.; Schotten, U.; Mahajan, R.; Antic, N.; Hatem, S.; Pathak, R.; Hendriks, J.; Kalman, J.; Sanders, P. Novel mechanisms in the pathogenesis of atrial fibrillation: Practical applications. *Eur. Heart J.* **2016**, *37*, 1573–1581. [CrossRef] [PubMed]

9. Allessie, M. Electrical, contractile and structural remodeling during atrial fibrillation. *Cardiovasc. Res.* **2002**, *54*, 230–246. [CrossRef]

10. Yamaguchi, T.; Marrouche, N.F. Recurrence Post-Atrial Fibrillation Ablation: Think Outside the Pulmonary Veins. *Circ. Arrhythmia Electrophysiol.* **2018**, *11*, e006379. [CrossRef] [PubMed]

11. Yoshida, K.; Ulfarsson, M.; Tada, H.; Chugh, A.; Good, E.; Kuhne, M.; Crawford, T.; Sarrazin, J.F.; Chalfoun, N.; Wells, D.; et al. Complex Electrograms Within the Coronary Sinus: Time- and Frequency-Domain Characteristics, Effects of Antral Pulmonary Vein Isolation, and Relationship to Clinical Outcome in Patients with Paroxysmal and Persistent Atrial Fibrillation. *J. Cardiovasc. Electrophysiol.* **2008**, *19*, 1017–1023. [CrossRef]

12. Nademanee, K.; McKenzie, J.; Kosar, E.; Schwab, M.; Sunsaneewitayakul, B.; Vasavakul, T.; Khunnawat, C.; Ngarmukos, T. A new approach for catheter ablation of atrial fibrillation: Mapping of the electrophysiologic substrate. *J. Am. Coll. Cardiol.* **2004**, *43*, 2044–2053. [CrossRef] [PubMed]

13. de Groot, N.; Shah, D.; Boyle, P.M.; Anter, E.; Clifford, G.D.; Deisenhofer, I.; Deneke, T.; van Dessel, P.; Doessel, O.; Dilaveris, P.; et al. Critical appraisal of technologies to assess electrical activity during atrial fibrillation: A position paper from the European Heart Rhythm Association and European Society of Cardiology Working Group on eCardiology in collaboration with the Heart Rhythm Society, Asia Pacific Heart Rhythm Society, Latin American Heart Rhythm Society and Computing in Cardiology. *EP Eur.* **2021**, *24*, 313–330. [CrossRef]

14. Verma, A.; Jiang, C.; Betts, T.; Chen, J.; Deisenhofer, I.; Mantovan, R.; Macle, L.; Morillo, C.; Haverkamp, W.; Weerasooriya, R.; et al. Approaches to catheter ablation for persistent atrial fibrillation. *N. Engl. J. Med.* **2015**, *372*, 1812–1822. [CrossRef] [PubMed]

15. Ammar-Busch, S.; Reents, T.; Knecht, S.; Rostock, T.; Arentz, T.; Duytschaever, M.; Neumann, T.; Cauchemez, B.; Albenque, J.P.; Hessling, G.; et al. Correlation between atrial fibrillation driver locations and complex fractionated atrial electrograms in patients with persistent atrial fibrillation. *Pacing Clin. Electrophysiol.* **2018**, *41*, 1279–1285. [CrossRef]

16. Maille, B.; Das, M.; Hussein, A.; Shaw, M.; Chaturvedi, V.; Williams, E.; Morgan, M.; Ronayne, C.; Snowdon, R.L.; Gupta, D. Reverse electrical and structural remodeling of the left atrium occurs early after pulmonary vein isolation for persistent atrial fibrillation. *J. Interv. Card. Electrophysiol.* **2020**, *58*, 9–19. [CrossRef] [PubMed]

17. Lau, D.; Linz, D.; Schotten, U.; Mahajan, R.; Sanders, P.; Kalman, J. Pathophysiology of Paroxysmal and Persistent Atrial Fibrillation: Rotors, Foci and Fibrosis. *Heart Lung Circ.* **2017**, *26*, 887–893. [CrossRef] [PubMed]

18. Botteron, G.; Smith, J. A Technique for Measurement of the Extent of Spatial Organization of Atrial Activation During Atrial Fibrillation in the Intact Human Heart. *IEEE Trans. Biomed. Eng.* **1995**, *42*, 579–586. [CrossRef]

19. Faes, L.; Nollo, G.; Antolini, R.; Gaita, F.; Ravelli, F. A method for quantifying atrial fibrillation organization based on wave-morphology similarity. *IEEE Trans. Biomed. Eng.* **2002**, *49*, 1504–1513. [CrossRef] [PubMed]

20. Houben, R.; Allessie, M. Processing of intracardiac electrograms in atrial fibrillation: Diagnosis of electropathological substrate of AF. *IEEE Eng. Med. Biol. Mag.* **2006**, *25*, 40–51. [CrossRef]

21. Lin, Y.J.; Lo, M.T.; Lin, C.; Chang, S.L.; Lo, L.W.; Hu, Y.F.; Chao, T.F.; Li, C.H.; Chang, Y.C.; Hsieh, W.H.; et al. Nonlinear analysis of fibrillatory electrogram similarity to optimize the detection of complex fractionated electrograms during persistent atrial fibrillation. *J. Cardiovasc. Electrophysiol.* **2013**, *24*, 280–289. [CrossRef]

22. Ng, J.; Sehgal, V.; Ng, J.; Gordon, D.; Goldberger, J. Iterative method to detect atrial activations and measure cycle length from electrograms during atrial fibrillation. *IEEE Trans. Biomed. Eng.* **2014**, *61*, 273–278. [CrossRef] [PubMed]

23. Dalvi, R.; Suszko, A.; Chauhan, V.S. Graph search based detection of periodic activations in complex periodic signals: Application in atrial fibrillation electrograms. In Proceedings of the 2015 IEEE 28th Canadian Conference on Electrical and Computer Engineering (CCECE), Halifax, NS, Canada, 3–6 May 2015; pp. 376–381. [CrossRef]

24. Osorio, D.; Vraka, A.; Quesada, A.; Hornero, F.; Alcaraz, R.; Rieta, J.J. An Efficient Hybrid Methodology for Local Activation Waves Detection under Complex Fractionated Atrial Electrograms of Atrial Fibrillation. *Sensors* **2022**, *22*, 5345. [CrossRef] [PubMed]

25. Costabal, F.S.; Zaman, J.A.B.; Kuhl, E.; Narayan, S.M. Interpreting Activation Mapping of Atrial Fibrillation: A Hybrid Computational/Physiological Study. *Ann. Biomed. Eng.* **2018**. *46*, 257–269. [CrossRef]

26. Abdi, B.; Hendriks, R.C.; van der Veen, A.J.; de Groot, N.M.S. Improved local activation time annotation of fractionated atrial electrograms for atrial mapping. *Comput. Biol. Med.* **2020**. *117*, 103590. [CrossRef]
27. Kölling, B.; Abdi, B.; de Groot, N.M.; Hendriks, R.C. Local Activation Time Estimation in Atrial Electrograms Using Cross-Correlation over Higher-Order Neighbors. In Proceedings of the 2020 28th European Signal Processing Conference (EUSIPCO), Dublin, Ireland, 23–27 August 2021; pp. 905–909. [CrossRef]
28. Wells, J.L., Jr.; Karp, R.; Kouchoukos, N.; MacLean, W.; James, T.; Waldo, A. Characterization of Atrial Fibrillation in Man: Studies Following Open Heart Surgery. *Pacing Clin. Electrophysiol.* **1978**, *1*, 426–438. [CrossRef] [PubMed]
29. Martínez-Iniesta, M.; Ródenas, J.; Alcaraz, R.; Rieta, J.J. Waveform Integrity in Atrial Fibrillation: The Forgotten Issue of Cardiac Electrophysiology. *Ann. Biomed. Eng.* **2017**, *45*, 1890–1907. [CrossRef]

Journal of
Personalized Medicine

Article

A Differential Profile of Biomarkers between Patients with Atrial Fibrillation and Healthy Controls

Ana Merino-Merino [1], Ruth Saez-Maleta [2], Ricardo Salgado-Aranda [1,3], Daniel AlKassam-Martinez [2,4], Virginia Pascual-Tejerina [1,5], Javier Martin-Gonzalez [1], Javier Garcia-Fernandez [1] and Jose-Angel Perez-Rivera [1,6,*]

[1] Cardiology Department, Hospital Universitario de Burgos, 09006 Burgos, Spain
[2] Clinical Analyses Department, Hospital Universitario de Burgos, 09006 Burgos, Spain
[3] Cardiology Department, Hospital Clínico San Carlos, 28040 Madrid, Spain
[4] Clinical Analyses Department, Hospital Central de Asturias, 33011 Oviedo, Asturias, Spain
[5] Cardiology Department, Hospital Virgen de la Salud, 45004 Toledo, Spain
[6] Universidad Isabel I, 09003 Burgos, Spain
* Correspondence: jangel.perezrivera@secardiologia.es; Tel.: +34-947281800 (ext. 35756)

Abstract: Atrial fibrillation (AF) is explained by anatomical and electrophysiological changes in the atria determined by high pressure, dilatation, infiltration and inflammation in the myocardium. There are some biomarkers implicated in these processes, namely, NT-proBNP, high sensitivity troponin (Hs-Tn), urate, galectin-3, ST2, C reactive protein and fibrinogen. The aim of this study was to assess differences in these biomarkers between patients with AF and healthy controls. We designed a cross-sectional study consecutively including all patients undergoing electrical cardioversion in our hospital for persistent AF and matched healthy controls. We included 115 patients with persistent non-valvular AF and 33 healthy subjects. The biomarkers NT-proBNP, ST2 and Hs-Tn T were significantly related to the presence of AF (1054 ± 833.30 vs. 58.31 ± 59.40, $p < 0.001$; 35.43 ± 15.89 vs. 27.43 ± 10.95, $p < 0.001$ and 10.25 ± 6.11 vs. 8.42 ± 6.85, $p < 0.001$, respectively). NT-proBNP was the best biomarker differentiating AF patients (area under the curve 0.995). The best NT-proBNP cut-off point to differentiate AF was 102 pg/mL; for Hs-Tn T it was 11.5 ng/L and for ST2 it was 37.7 ng/mL. It is possible that these biomarkers intervene at the onset of AF and have no role in AF maintenance.

Keywords: atrial fibrillation; biomarkers; NT-proBNP

Citation: Merino-Merino, A.; Saez-Maleta, R.; Salgado-Aranda, R.; AlKassam-Martinez, D.; Pascual-Tejerina, V.; Martin-Gonzalez, J.; Garcia-Fernandez, J.; Perez-Rivera, J.-A. A Differential Profile of Biomarkers between Patients with Atrial Fibrillation and Healthy Controls. *J. Pers. Med.* **2022**, *12*, 1406. https://doi.org/10.3390/jpm12091406

Academic Editor: José Miguel Rivera-Caravaca

Received: 3 August 2022
Accepted: 26 August 2022
Published: 30 August 2022

1. Introduction

Atrial fibrillation (AF) is the most frequent arrhythmia. It is explained by anatomical and electrophysiological changes in the atrial myocardium determined by high pressure, dilatation, infiltration and inflammation. Identification of biomarkers associated with AF may advance knowledge of AF, increasing our understanding of the pathophysiological mechanisms of the arrhythmia. Including biomarkers associated with AF in risk scales may yield predictions of AF risk with more precision in the future. Moreover, biomarkers may be used in the development of pharmacological pathways for AF preventive therapies [1].

Some biomarkers are implicated in processes involved in AF onset and progression, namely, NT-proBNP (implicated in mechanical stress and myocardial stretch), high sensitivity troponin T (Hs-Tn T) (a myocardial damage biomarker), urate (associated with oxidative stress), galectin-3 and ST2 (implicated in remodeling and fibrosis), C reactive protein (CRP) and fibrinogen (implicated in inflammation) [2–4].

The relationship between biomarkers and AF has been previously shown [5], but the potential value of combining several biomarkers to achieve an integrated assessment is still not fully established [6].

The aim of this study was to assess differences in these biomarkers between AF patients and healthy subjects.

2. Materials and Methods

2.1. Design and Population

This cross-sectional study included all consecutive stable patients presenting with non-valvular persistent AF, non-urgently submitted to our unit for electrical cardioversion between 17 April 2015 and 14 July 2017. We selected 1:3 aged-paired controls.

An echocardiogram performed at our Cardiac Image Unit in the 6 months prior to the electrical cardioversion was required (main echocardiographic measurements are shown in Supplementary Table S1). We excluded patients with significant structural cardiac abnormalities (moderate or severe valvular disease, valvular prosthesis, history of LVEF less than 40%, hypertrophic cardiomyopathy and infiltrative cardiomyopathy), presence of atrial flutter or arrhythmias other than AF, previous cardioversion or pulmonary vein ablation, patients with clinical instability and asymptomatic patients. None of the controls had a history of AF or any other cardiovascular disease in any of its forms.

In the basal clinical interview, a cardiologist checked the inclusion and exclusion criteria and recruited patients and controls agreeing to sign the informed consent form. The local Ethics Committee's approval was obtained for this study (reference number, CEIC-1407). The study was performed in accordance with the Declaration of Helsinki. Prior to cardioversion, blood samples of all patients were obtained. Blood samples of the controls were obtained on the day of inclusion. The biomarkers determined by our center's laboratory were NT-proBNP, Hs-Tn T, galectin-3, ST2, fibrinogen, urate and CRP. Glomerular filtration rates were estimated using the CKD-EPI equation.

2.2. Statistical Analysis

SPSS version 20.0 for Windows (IBM, Chicago, IL, USA) was used to perform the statistical analysis. Quantitative variables were expressed as mean and standard deviation or medians, and as interquartile ranges when a normal distribution was not observed, as per the Kolmogorov–Smirnov goodness-of-fit test. Qualitative variables were expressed as frequency and percentage. In order to assess differences in biomarker levels and clinical variables in subjects with or without AF, a univariate analysis was performed using the *t*-test for normal quantitative variables, the U–Mann–Witney test for non-normal quantitative variables, and the Chi-square test for qualitative variables. Receiver operating characteristic (ROC) curves were obtained to assess the biomarkers' most accurate diagnostic cut-off values. The best cut-off point corresponded to the maximum vertical distance between the ROC curve and the diagonal line. Finally, we performed multivariate analysis using logistic regression analysis including those variables that showed statistical significance ($p < 0.05$) in the univariate analysis.

3. Results

3.1. Clinical Differences between Cases and Controls

We included 115 patients with AF and 33 healthy controls. Differences in clinical characteristics between cases and controls are shown in Table 1. Only the male sex was significantly related to the presence of AF (71.30% vs. 51.51%; $p = 0.033$).

Table 1. Clinical differences between cases and controls.

	Cases (N = 115)	Controls (N = 33)	*p*
Age	63 ± 9	62 ± 10	0.464
Men	82 (71.30%)	17 (51.51%)	0.033
Hypertension	65 (56.52%)	15 (45.45%)	0.261
Diabetes	15 (13.04%)	4 (12.12%)	0.889
COPD	5 (4.34%)	1 (3.03%)	0.729
Smoking	19 (16.52%)	8 (24.24%)	0.322
CRD	2 (1.739%)	1 (3.03%)	0.643
Stroke	3 (2.60%)	1 (3.03%)	0.895
Previous myocardial infarction	6 (5.21%)	1 (3.03%)	0.602
OSAHS	10 (8.69%)	2 (6.06%)	0.625

CRD: Chronic Renal Dysfunction; COPD: Chronic Obstructive Pulmonary Disease.

3.2. Analytical Differences between Cases and Controls

Differences in analytical characteristics (including biomarkers) between cases and controls observed in the univariate analysis are shown in Table 2. Renal function, measured as creatinine levels and glomerular filtration (0.95 ± 0.19 mg/dl vs. 0.79 ± 0.12 mg/dl; $p < 0.001$ and 79.45 ± 15.23 mL/min vs. 90.54 ± 10.07 mL/min; $p < 0.001$, respectively), was significantly related to the presence of AF. Biomarkers NT-proBNP (1054.20 ± 833.30 pg/mL vs. 58.31 ± 59.40 pg/mL; $p < 0.001$), ST2 (35.43 ± 15.89 ng/mL vs. 27.43 ± 10.95 ng/mL; $p < 0.001$) and Hs-Tn T (10.25 ± 6.11 ng/L vs. 8.42 ± 6.85 ng/L; $p < 0.001$) were also significantly related to the presence of AF.

Table 2. Analytical differences between cases and controls.

	Cases (N = 115)	Controls (N = 33)	*p*
NT-proBNP (pg/mL)	1054.20 ± 833.30	58.31 ± 59.40	<0.001
Galectin-3 (ng/mL)	16.87 ± 4.89	22.71 ± 21.94	0.139
ST2 (ng/mL)	35.43 ± 15.89	27.43 ± 10.95	<0.001
Fibrinogen (mg/dl)	329.40 ± 75.87	315.33 ± 73.31	0.346
Hs-Tn T (ng/L)	10.25 ± 6.11	8.42 ± 6.85	<0.001
Urate (mg/dl)	6.11 ± 1.45	6.38 ± 7.68	0.845
CRP (mg/L)	5.06 ± 14.8	2.46 ± 2.10	0.326
Hemoglobin (g/dl)	14.97 ± 1.43	14.56 ± 1.29	0.150
Creatinine (mg/dl)	0.95 ± 0.19	0.79 ± 0.12	<0.001
Glomerular filtration (mL/min)	79.45 ± 15.23	90.54 ± 10.07	<0.001

CRP: C-reactive protein; Hs-Tn T: High sensitivity troponin T.

3.3. ROC Test

To assess the biomarkers' yield, we performed ROC tests including those biomarkers that showed a significant relationship with the presence of AF in the univariate analysis. The area under the ROC curve for NT-proBNP was 0.995, for Hs-Tn T it was 0.655 and for ST2 it was 0.648. ROC curves are shown in Figure 1.

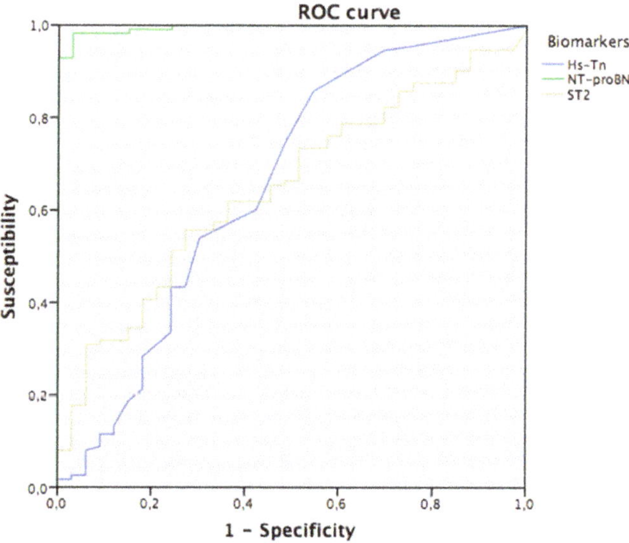

Figure 1. ROC curves. Hs-Tn: High sensitivity troponin.

We used the deLong test to compare the AUCs of three biomarkers. NT-proBNP was significantly better than Hs-Tn T ($p < 0.001$) and better than ST2 ($p < 0.001$), but there were no significant differences between Hs-Tn T and ST2 ($p = 0.99$). NT-proBNP was the best biomarker for differentiating patients with AF.

From this test, we obtained the best cut-off points to differentiate cases and controls for three biomarkers. The best cut-off point for NT-proBNP was 102 pg/mL (99% sensibility and 76% specificity), for Hs-Tn T it was 11.5 ng/L (28% sensibility and 82% specificity) and for ST2 it was 37.7 ng/mL (40% sensitivity and 82% specificity).

3.4. Multivariate Analysis

To perform multivariate analysis we included those clinical variables and biomarkers that showed a statistical significance in the univariate analysis: male sex, glomerular filtration (we included this instead of creatinine because it provides information regarding renal function that is more accurate) and biomarkers Hs-Tn T, NT-proBNP and ST2. We present the results with no dichotomized biomarkers levels, and then with dichotomized biomarkers levels, using the best cut-off points from the ROC curve.

In the first analysis (Table 3), NT-proBNP was the only variable independently related to the presence of AF (odds ratio 1.03; 95% confidence interval 1.01–1.04; $p < 0.001$).

Table 3. Multivariate analysis.

	Odds Ratio	95% Confidence Interval	p
NT-proBNP (pg/mL)	1.03	1.01–1.04	<0.001
ST2 (ng/mL)	1.25	0.88–1.79	0.215
Hs-Tn T (ng/L)	0.81	0.61–1.08	0.159
Men	37.60	0.51–2770.86	0.098
Glomerular filtration (mL/min)	0.93	0.82–1.04	0.220

Hs-Tn T: High sensitivity troponin T.

In the second analysis, NT-proBNP was the only variable independently related to the presence of AF (odds ratio 442.16; 95% confidence interval 46.27–4224.83; $p < 0.001$). The odds ratio in this case was remarkably high. We found two possibilities that explain this, including the small sample size and that the vast majority of patients (97.39%) showed NT-proBNP levels above the cut-off point. As a result, in samples such as ours, the odds ratio is particularly high and should not be taken into account.

4. Discussion

In our sample, we detected higher levels of cardiac biomarkers in AF patients than in healthy controls. NT-proBNP showed the best performance in discriminating cases and controls.

We found a higher proportion of men in the patient group with AF than in healthy controls (71.30% vs. 51.51%; $p = 0.033$). According to previous studies, this might be due to the high prevalence of cardiovascular risk factors among men [7]. We paired controls by age, but not by sex; therefore, differences may have been found by chance. Nevertheless, we included sex in the multivariate analysis to avoid confounding factors.

NT-proBNP, ST2 and Hs-Tn T were the only biomarkers significantly related to the presence of AF. Similar to our study, NT-proBNP has previously been independently related to the presence of AF [8,9]. Moreover, it has been demonstrated that the development and progression of AF (from paroxysmal to persistent) is associated with a gradual increase in NT-proBNP levels [10].

Although NT-proBNP levels are related to a higher risk of AF, cut-off points and treatments based on those points are not yet established. In our study, the best NT-proBNP cut-off point was 102 pg/mL (99% sensitivity and 76% specificity). Palà et al. [11] noted a similar cut-off point for NT-proBNP (95 pg/mL, 95% sensibility and 66.2% specificity). The association between NT-proBNP and the presence of AF can be explained by atria remodeling (in which NT-proBNP is implicated) when it is expressed secondary to atrial distension and dilatation. Our research group has previously shown a relationship between NT-proBNP and recurrences of AF. Patients with persistent high values of this biomarker have active processes of atrial stretch, remodeling and fibrosis; these mechanisms are probably the most important contributors to AF maintenance. Furthermore, it seems that a new onset of AF can reactivate inflammation and fibrosis in the acute phase, and over the time this mechanism might decrease [12].

ST2 has previously been related to AF and has been shown to have higher values in patients with persistent AF compared to patients with paroxysmal AF, which can translate progression of the AF [13]. In contrast, ST2 values were significantly higher in patients with persistent and permanent AF compared to patients in sinus rhythm; however, no significant differences were found between persistent and permanent AF [14].

In our study, the best ST2 cut-off point to discriminate AF was 37.7 ng/mL (40% sensitivity and 82% specificity). The association between ST2 and AF can be explained by the implication of ST2 in fibrosis and remodeling processes that initiate and maintain AF. A performance algorithm has been described in maintaining sinus rhythm based on ST2 values [15]. This algorithm was based on the hypothesis that elevated ST2 levels translate into excess myocardial fibrosis. Therefore, patients with high ST2 levels (considering the cut-off point of 35 ng/mL) would not benefit from performing electrical cardioversion and should be evaluated in a specialized consultation to assess pulmonary vein ablation [16].

Hs-Tn T has not been classically related to AF, but it has been shown that high levels of this biomarker are associated with the incidence of AF [17]. Increased levels of Hs-Tn T in AF patients in our study was probably a result of the myocyte damage that can occur in AF.

In our study, the best Hs-Tn T cut-off point to discriminate AF was 11.5 n/L (28% sensitivity and 82% specificity). A relationship between Hs-Tn and AF has been previously shown. A metanalysis including 27 studies showed significantly higher Hs-Tn levels in AF patients than in subjects without AF [18].

In a recent study that included more than 3000 patients with mild or moderate chronic kidney disease with a 7-year follow-up, Hs-Tn T values were associated with a higher risk of AF onset [19]. Another study with 241 AF patients and 824 subjects with no cardiovascular disease showed increased Hs-Tn T levels in those with AF. Moreover, patients with persistent AF showed higher levels of Hs-Tn T than those with paroxysmal AF. That study also showed a relationship between Hs-Tn T and the presence of low voltage areas in the left atria. The authors explained that these areas translate remodeling and fibrosis zones with proapoptotic processes; however, the progression of AF does not necessarily translate the destruction of cardiomyocytes [20].

In a multicenter study of hospitalized patients with COVID-19, troponin and NT-proBNP levels were significantly higher in patients with a history of AF than in patients without a history of AF. Nevertheless, there were no differences in other biomarkers, such as CRP [21].

Several biomarkers (urate, galectin-3, fibrinogen and CRP) did not show any relationship with the presence of AF in our sample. Our AF patients presented different AF durations, but they all had persistent AF. It is possible that these biomarkers intervene at the onset of AF and have no role in AF maintenance; it can also be explained by the small sample size.

Regarding the rest of the analytical values, creatinine levels were higher in AF cases than in controls and glomerular filtration were lower in AF cases than in controls. A bidirectional relationship exists between kidney disease and cardiovascular disease (including AF) [22]. In fact, chronic kidney disease is a predictor of cardiovascular disease as well as the onset of AF; it is two or three times more likely than in patients without chronic kidney disease [23]. On the other hand, the presence of AF is related to the progression of kidney disease [24].

Our study has some limitations. We only included patients with symptomatic persistent AF; therefore, our results cannot be extrapolated to other populations. We did not take differences in AF duration between our patients into account. We did not analyze different variables, such as previous electrical cardioversion or pulmonary veins ablation, which could impact biomarker levels and might also be affected by the presence of AF. Finally, given our study design, we cannot elucidate whether the increase in biomarker levels in our study's sample population was a cause or a consequence of AF.

On the other hand, to our knowledge this is the first study analyzing a wide battery of biomarkers in AF patients and healthy controls. The identification of pathophysiological phenomena of atrial remodeling could be useful in detecting individuals at risk of developing AF. Using biomarkers to detect AF risk could facilitate the application of more exhaustive diagnostic procedures for affected patients.

5. Conclusions

In our sample, NT-proBNP, ST2 and Hs-Tn T were related to the presence of AF. NT-proBNP showed the highest yield in differentiating patients with AF from healthy subjects, and was the only biomarker independently related to the presence of AF.

Supplementary Materials: The following supporting information can be downloaded at: https://www.mdpi.com/article/10.3390/jpm12091406/s1, Table S1: Echocardiographic measurements in patients with atrial fibrillation.

Author Contributions: Conceptualization, J.-A.P.-R.; methodology, J.-A.P.-R. and J.G.-F.; software, V.P.-T. and J.M.-G.; validation, J.-A.P.-R., J.G.-F. and R.S.-M.; formal analysis, D.A.-M. and R.S.-M.; investigation, V.P.-T. and A.M.-M.; resources, J.M.-G., J.G.-F., R.S.-A., D.A.-M. and R.S.-M.; data curation, A.M.-M., V.P.-T. and J.-A.P.-R.; writing—original draft preparation, A.M.-M.; writing—review and editing, R.S.-A. and J.-A.P.-R.; visualization, R.S.-A.; supervision, J.-A.P.-R.; project administration, J.-A.P.-R.; funding acquisition, J.-A.P.-R. All authors have read and agreed to the published version of the manuscript.

Funding: This research was funded by the Regional Health Authority of Castilla y Leon [grant numbers GRS 1114/A/15 and INT/M/04/22 to Dr. Perez-Rivera].

Institutional Review Board Statement: The local Ethics Committee's approval was obtained for this study (reference number, CEIC-1407). The study was performed in accordance with the Declaration of Helsinki.

Informed Consent Statement: Informed consent was obtained from all subjects involved in this study.

Data Availability Statement: Raw data were generated at the University Hospital of Burgos. Derived data supporting the findings of this study are available from the corresponding author, J.-A.P.-R., on request.

Conflicts of Interest: The authors declare no conflict of interest.

References

1. Staerk, L.; Preis, S.R.; Lin, H.; Lubitz, S.A.; Ellinor, P.T.; Levy, D.; Benjamin, E.J.; Trinquart, L. Protein Biomarkers and Risk of Atrial Fibrillation: The FHS. *Circ. Arrhythm. Electrophysiol.* **2020**, *13*, e007607. [CrossRef] [PubMed]
2. Gottdiener, J.S.; Seliger, S.; de Filippi, C.; Christenson, R.; Baldridge, A.S.; Kizer, J.R.; Psaty, B.M.; Shah, S.J. Relation of Biomarkers of Cardiac Injury, Stress, and Fibrosis with Cardiac Mechanics in Patients ≥ 65 Years of Age. *Am. J. Cardiol.* **2020**, *136*, 156–163. [CrossRef] [PubMed]
3. Kwon, C.H.; Kang, J.G.; Lee, H.J.; Kim, N.H.; Sung, J.-W.; Cheong, E.; Sung, K.-C. C-Reactive Protein and Risk of Atrial Fibrillation in East Asians. *EP Eur.* **2017**, *19*, 1643–1649. [CrossRef] [PubMed]
4. Canpolat, U.; Aytemir, K.; Yorgun, H.; Şahiner, L.; Kaya, E.B.; Çay, S.; Topaloğlu, S.; Aras, D.; Oto, A. Usefulness of Serum Uric Acid Level to Predict Atrial Fibrillation Recurrence after Cryoballoon-Based Catheter Ablation. *EP Eur.* **2014**, *16*, 1731–1737. [CrossRef]
5. O'Neal, W.T.; Venkatesh, S.; Broughton, S.T.; Griffin, W.F.; Soliman, E.Z. Biomarkers and the prediction of atrial fibrillation: State of the art. *Vasc. Health Risk Manag.* **2016**, *12*, 297–303. [CrossRef]
6. Boriani, G.; Valenti, A.C.; Vitolo, M. Biomarkers in atrial fibrillation: A constant search for simplicity, practicality, and cost-effectiveness. *Kardiol. Pol.* **2021**, *79*, 243–245. [CrossRef]
7. Gullón, P.; Díez, J.; Cainzos-Achirica, M.; Franco, M.; Bilal, U. Social inequities in cardiovascular risk factors in women and men by autonomous regions in Spain. *Gac. Sanit.* **2021**, *35*, 326–332. [CrossRef]
8. Dudink, E.A.; Weijs, B.; Tull, S.; Luermans, J.G.; Fabritz, L.; Chua, W.; Rienstra, M.; Gelder, I.C.V.; Schotten, U.; Kirchhof, P.; et al. The Biomarkers NT-proBNP and CA-125 are Elevated in Patients with Idiopathic Atrial Fibrillation. *J. Atr. Fibrillation* **2018**, *11*, 2058. [CrossRef]

9. Wang, T.J.; Larson, M.G.; Levy, D.; Benjamin, E.J.; Leip, E.P.; Omland, T.; Wolf, P.A.; Vasan, R.S. Plasma natriuretic peptide levels and the risk of cardiovascular events and death. *N. Engl. J. Med.* **2004**, *350*, 655–663. [CrossRef]

10. Stanciu, A.E.; Vatasescu, R.G.; Stanciu, M.M.; Serdarevic, N.; Dorobantu, M. The role of pro-fibrotic biomarkers in paroxysmal and persistent atrial fibrillation. *Cytokine* **2018**, *103*, 63–68. [CrossRef]

11. Palà, E.; Bustamante, A.; Clúa-Espuny, J.L.; Acosta, J.; Gonzalez-Loyola, F.; Ballesta-Ors, J.; Gill, N.; Caballero, A.; Pagola, J.; Pedrote, A.; et al. N-Terminal Pro B-Type Natriuretic Peptide's Usefulness for Paroxysmal Atrial Fibrillation Detection among Populations Carrying Cardiovascular Risk Factors. *Front. Neurol.* **2019**, *10*, 1226. [CrossRef] [PubMed]

12. Merino-Merino, A.; Saez-Maleta, R.; Salgado-Aranda, R.; AlKassam-Martinez, D.; Pascual-Tejerina, V.; Martin-González, J.; Garcia-Fernandez, J.; Perez-Rivera, J.-A. When should we measure biomarkers in patients with atrial fibrillation to predict recurrences? *Am. J. Emerg. Med.* **2021**, *39*, 248–249. [CrossRef] [PubMed]

13. Ma, X.; Yuan, H.; Luan, H.-X.; Shi, Y.-L.; Zeng, X.-L.; Wang, Y. Elevated soluble ST2 concentration may involve in the progression of atrial fibrillation. *Clin. Chim. Acta* **2018**, *480*, 138–142. [CrossRef] [PubMed]

14. Chen, C.; Qu, X.; Gao, Z.; Zheng, G.; Wang, Y.; Chen, X.; Li, H.; Huang, W.; Zhou, H. Soluble ST2 in Patients with Nonvalvular Atrial Fibrillation and Prediction of Heart Failure. *Int. Heart J.* **2018**, *59*, 58–63. [CrossRef]

15. Chang, K.-W.; Hsu, J.C.; Toomu, A.; Fox, S.; Maisel, A.S. Clinical Applications of Biomarkers in Atrial Fibrillation. *Am. J. Med.* **2017**, *130*, 1351–1357. [CrossRef]

16. Merino-Merino, A.; Gonzalez-Bernal, J.; Fernandez-Zoppino, D.; Saez-Maleta, R.; Perez-Rivera, J.-A. The Role of Galectin-3 and ST2 in Cardiology: A Short Review. *Biomolecules* **2021**, *11*, 1167. [CrossRef]

17. Filion, K.B.; Agarwal, S.K.; Ballantyne, C.M.; Eberg, M.; Hoogeveen, R.C.; Huxley, R.R.; Loehr, L.R.; Nambi, V.; Soliman, E.Z.; Alonso, A. High-sensitivity cardiac troponin T and the risk of incident atrial fibrillation: The Atherosclerosis Risk in Communities (ARIC) study. *Am. Heart J.* **2015**, *169*, 31–38.e3. [CrossRef]

18. Bai, Y.; Guo, S.-D.; Liu, Y.; Ma, C.-S.; Lip, G.Y.H. Relationship of troponin to incident atrial fibrillation occurrence, recurrence after radiofrequency ablation and prognosis: A systematic review, meta-analysis and meta-regression. *Biomarkers* **2018**, *23*, 512–517. [CrossRef]

19. Janus, S.E.; Hajjari, J.; Al-Kindi, S. High-sensitivity troponin and the risk of atrial fibrillation in chronic kidney disease: Results from the Chronic Renal Insufficiency Cohort Study. *Heart Rhythm* **2020**, *17*, 190–194. [CrossRef]

20. Kornej, J.; Zeynalova, S.; Büttner, P.; Burkhardt, R.; Bae, Y.J.; Willenberg, A.; Baber, R.; Thaler, A.; Hindricks, G.; Loeffler, M.; et al. Differentiation of atrial fibrillation progression phenotypes using Troponin T. *Int. J. Cardiol.* **2019**, *297*, 61–65. [CrossRef]

21. Paris, S.; Inciardi, R.M.; Lombardi, C.M.; Tomasoni, D.; Ameri, P.; Carubelli, V.; Agostoni, P.; Canale, C.; Carugo, S.; Danzi, G.; et al. Implications of atrial fibrillation on the clinical course and outcomes of hospitalized COVID-19 patients: Results of the Cardio-COVID-Italy multicentre study. *Europace* **2021**, *23*, 1603–1611. [CrossRef] [PubMed]

22. Miyazawa, K.; Pastori, D.; Lip, G.Y.H. Changes in renal function in patients with atrial fibrillation: Efficacy and safety of the non-vitamin K antagonist oral anticoagulants. *Am. Heart J.* **2018**, *198*, 166–168. [CrossRef] [PubMed]

23. Alonso, A.; Lopez, F.L.; Matsushita, K.; Loehr, L.R.; Agarwal, S.K.; Chen, L.Y.; Soliman, E.Z.; Astor, B.C.; Coresh, J. Chronic kidney disease is associated with the incidence of atrial fibrillation: The Atherosclerosis Risk in Communities (ARIC) study. *Circulation* **2011**, *123*, 2946–2953. [CrossRef] [PubMed]

24. Watanabe, H.; Watanabe, T.; Sasaki, S.; Nagai, K.; Roden, D.M.; Aizawa, Y. Close bidirectional relationship between chronic kidney disease and atrial fibrillation: The Niigata preventive medicine study. *Am. Heart J.* **2009**, *158*, 629–636. [CrossRef] [PubMed]

Journal of
Personalized Medicine

Article

Inversion of Left Atrial Appendage Will Cause Compressive Stresses in the Tissue: Simulation Study of Potential Therapy

Salvatore Pasta [1], Julius M. Guccione [2] and Ghassan S. Kassab [3,*]

[1] Department of Engineering, Viale delle Scienze, Università degli Studi di Palermo, 90128 Palermo, Italy; salvatore.pasta@unipa.it
[2] Department of Surgery, University of California San Francisco, San Francisco, CA 94143, USA; julius.guccione@ucsf.edu
[3] California Medical Innovations Institute, 11107 Roselle, San Diego, CA 92121, USA
[*] Correspondence: gkassab@calmi2.org

Abstract: In atrial fibrillation (AF), thromboembolic events can result from the particular conformation of the left atrial appendage (LAA) bearing increased clot formation and accumulation. Current therapies to reduce the risk of adverse events rely on surgical exclusion or percutaneous occlusion, each of which has drawbacks limiting application and efficacy. We sought to quantify the hemodynamic and structural loads of a novel potential procedure to partially invert the "dead" LAA space to eliminate the auricle apex where clots develop. A realistic left atrial geometry was first achieved from the heart anatomy of the Living Heart Human Model (LHHM) and then the left atrial appendage inversion (LAAI) was simulated by finite-element analysis. The LAAI procedure was simulated by pulling the elements at the LAA tip and prescribing a displacement motion along a predefined path. The deformed configuration was then used to develop a computational flow analysis of LAAI. Results demonstrated that the inverted LAA wall undergoes a change in the stress distribution from tensile to compressive in the inverted appendage, and this can lead to resorption of the LAA tissue as per a reduced stress/resorption relationship. Computational flow analyses highlighted a slightly nested low-flow velocity pattern for the inverted LAA with minimal differences from that of a model without inversion of the LAA apex. Our study revealed important insights into the biomechanics of LAAI and demonstrated the inversion of the stress field (from tensile to compressive), which &can ultimately lead the long-term resorption of the LAA.

Keywords: finite element method; atrial fibrillation; atrophy; fibrosis

Citation: Pasta, S.; Guccione, J.M.; Kassab, G.S. Inversion of Left Atrial Appendage Will Cause Compressive Stresses in the Tissue: Simulation Study of Potential Therapy. *J. Pers. Med.* **2022**, *12*, 883. https://doi.org/10.3390/jpm12060883

Academic Editor: Chien-Hung Lee

Received: 4 May 2022
Accepted: 26 May 2022
Published: 27 May 2022

1. Introduction

Atrial fibrillation (AF), the most common sustained arrhythmia, affects 3–6 million Americans and increases the risk of stroke by 4 to 6 times on average [1,2]. AF prevalence and disease burden increase with age, accounting for 15% of all strokes, and with greater associated morbidity and mortality than non-AF related strokes [3]. In people >80 years old, AF is the direct cause of 1 in 4 strokes [1]. AF and related disorders have high individual and societal costs, ~$26 B per year in the US [4], and incidence is projected to more than double by 2035 [5].

The left atrial appendage (LAA) extends from the LA and creates a side chamber, which can be a site of increased clot formation and accumulation in AF. The LAA in a low-flow state as in AF is the nidus for >90% of thrombus formation [6,7], where the rapid contraction of the heart that accompanies AF can initiate the release of emboli and the consequent risk of stroke. Although the risk of thromboembolic events is reduced with long-term oral anticoagulation therapy, it is contraindicated in 7.8% of newly diagnosed AF patients [8], and only 50–60% of eligible patients with AF receive it [9]. Percutaneous [10–12] and surgical strategies [13,14] to occlude or exclude the LAA have been developed to reduce the risk of thromboembolic events, but current devices have major complications (e.g., perforation, migration, and incomplete closure) and disadvantages (e.g., high-cost and foreign body retention). Compared to endocardial occlusion, epicardial clipping or exclusion seems to have better hemodynamic and neurohormonal effects, but it is technically more difficult to achieve [12,15,16].

In this study, the feasibility of resorption of the "dead" LAA space without the need of a specific device left behind was investigated by computational modeling. Specifically, the procedure for the treatment of LAA uses a suction-based catheter device to latch the distal portion of the LAA endocardium (Figure 1), enabling partial LAA inversion (LAAI) as shown previously by our group [17]. The fundamental hypothesis is that the partial inversion of the LAA changes the stress distribution (from tensile to compressive) in the inverted appendage. An increase in tensile stress is known to cause tissue growth (e.g., hypertrophy in hypertension), whereas a decrease will cause resorption (e.g., atrophy in hypotension) as per a stress-growth law [18]. To test this hypothesis, the LAAI procedure was simulated using the realistic and accurate heart anatomy of the Living Heart Human Model® (LHHM) [19]. Then, the fluid flow circulating in the inverted appendage was assessed by computational-fluid dynamics and compared to that of LAA. Several structural and hemodynamic parameters were extrapolated to assess the feasibility of the LAAI procedure.

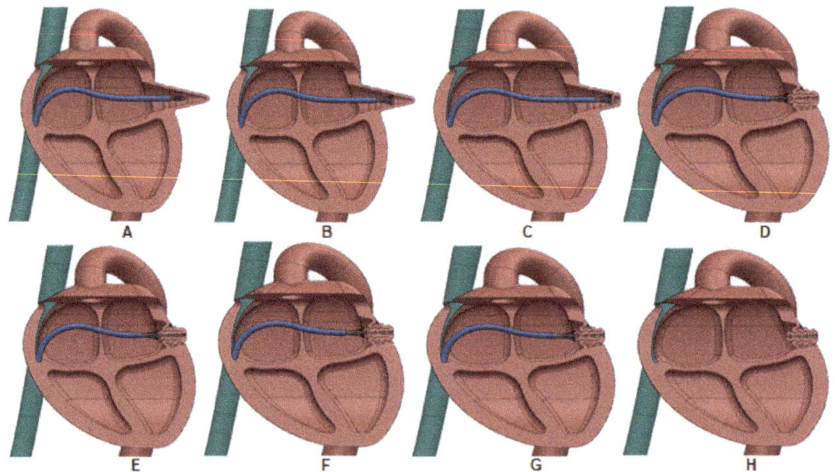

Figure 1. Schematic of LAA closure device and approach. The progression of the procedure is from (**A–H**). (**A**): Insert device into the LAA. (**B**): Deploy suction flute. (**C**): Suction onto inside distal wall of the LAA. (**D**): Pull the LAA inward and invert it. (**E**): Extend needle into inverted space. (**F**) Inject adhesive through needle-tip hypo-tube wire with side holes (blind closed tip). (**G**): Retract needle and maintain suction until adhesive cures. (**H**): Remove closure device leaving only tissue glue behind. The inversion of the LAA apex is not to scale, and can be much less, depending on the length of the LAA.

2. Materials and Methods

2.1. LAAI Structural Analysis

The geometry of the left atrium (LA) including the ear-shaped sac of the LAA was extrapolated from that of the LHHM representing the ideal average heart anatomy of a middle-aged healthy male. No human data was used in this study, and thus no authorization is needed by ethical committee. The LA geometry was then discretized with 33,632 triangular elements (S3) as a shell layer with uniform thickness of 2 mm [20]. A nearly incompressible material with an infinitesimal value of D and material density of $\rho = 1.06 \times 10^{-9}$ kg/mm^3 was used. Moreover, the biomechanical behavior of the LA tissue wall was modeled as a hyper-elastic and isotropic material using the third-order Ogden's strain energy function:

$$W \ = \ \sum_{i\,=\,1}^{3} \frac{\mu_i}{\alpha_i}(\lambda_1^{\alpha_i} \ + \ \lambda_2^{\alpha_i} \ + \ \lambda_3^{\alpha_i} \ - \ 3) \tag{1}$$

where λ_1 are the principle stretches with three pairs of material parameters μ and α. Specifically, material descriptors were $\mu_1 = -56.13$ MPa, $\alpha_1 = 8.65$, $\mu_2 = 42.88$ MPa, $\alpha_2 = 10.03$, $\mu_3 = 13.59$ MPa and $\alpha_1 = 6.82$ as obtained from the fitting of the passive myocardium [20].

To model the LAAI procedure, we first simulated the catheter clamping of the LAA by constraining the element nodes at the distal apex of the LAA. Then, the inversion was simulated pulling the clamped elements inside the heart by prescribing a displacement motion along a predefined path (i.e., the LAA centerline). Specifically, the clamped nodes were coupled to the centerline end point of LAA while the motion was implemented though connector assignments of constrained nodes in ABAQUS commercial software (ABAQUS v2020, Dassault Systèmes, Waltham, MA, USA). For each time increment, the displacement was 0.03 mm and was kept upon LAA inversion. To account for blood pressure, a uniform pressure of 1.3 mmHg was applied to LA inner surface. For boundary conditions, the distal ends of the pulmonary veins were fixed in all directions. The general contact algorithm with frictionless condition was adopted to consider the interaction of LAA tissue wall with itself during the retraction process. Simulations were carried out using ABAQUS/Explicit solver to account for the non-linear problem with large deformation and complex contact conditions. Energy was monitored to ensure the ratio of kinetic energy to internal energy remained less than 10%, and a variable mass-scaling technique was adopted to keep the time step less than 10^{-6}. Post-processing was carried out with Ensight software by overlaying the geometry of the whole LHHM heart.

2.2. LAAI Fluid Dynamic Analysis

Once simulation of LAAI was accomplished, the deformed configuration of the LAA was exported to generate the fluid domain and thus assess the left heart hemodynamic environment by unsteady computational flow analysis. Specifically, the inner volume of the LAA wall was meshed with 2,411,143 tetrahedral elements using the ICEM CFD (v21.0, ANSYS Inc., Canonsburg, PA, USA). Mesh quality check was evaluated by grid-convergence index analysis to quantify the discretization error using the pressure gradient on the LAA wall as convergence parameter, as previously [21]. The blood flow was assumed as laminar and incompressible with non-Newtonian viscosity described by the Carreau model [22]. The Navier–Stokes equations governing fluid motion were solved with an implicit algorithm in FLUENT (v21, ANSYS Inc., Canonsburg, PA, USA), which has been previously used to resolve high-frequency, time-dependent flow instabilities encountered in complex cardiovascular anatomies [23,24]. Pressure-implicit with splitting of operators (PISO) and skewness correction as pressure–velocity coupling, along with a pressure staggering option (PRESTO) scheme as pressure interpolation method and with second order accurate discretization was adopted. Convergence was enforced by reducing the residual of the continuity equation by 10^{-6} at every time step. For boundary conditions, the LAA wall was rigid with no-slip condition while flow velocity inlet (i.e., pulmonary veins) and outlet (mitral valve) with zero-pressure condition were set as flow conditions. For mitral valve outflow, a representative flow waveform with duration of 0.8 s at the mitral valve section was considered [25]. Afterwards, the inflow conditions were achieved by splitting the mitral outflow with a criterion based on proportionality of each pulmonary vein cross-sectional area on the basis of mass balance conservation. Each inlet and outflow were extended four times to ensure a fully developed flow at the entrance. To reduce the effect of transient flow, simulations were continued for three cardiac beats with the last cycle used for flow evaluation. The optimal solution was found for a time step of 0.02 s (i.e.,

400 steps for each cardiac cycle). For comparison, a reference model using the undeformed LAA configuration was also developed.

3. Results

Figure 2 shows the left atrial model with the LAA mesh at both undeformed and deformed configurations. The inversion of LAA was successfully simulated upon the neck region of the appendage tissue wall.

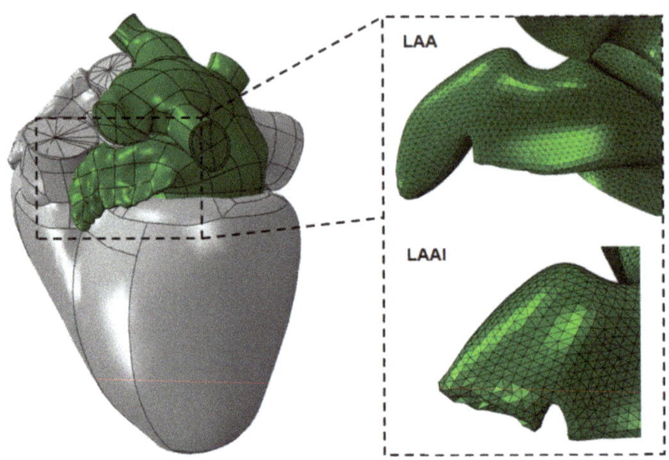

Figure 2. LA model (green) as extrapolated from the LHHM (grey); insertions show LAA mesh at undeformed and deformed configurations.

Figure 3 displays the map of the circumferential stress of LAA tissue wall at different steps of the inversion procedure. Compressive stress occurred in correspondence of the LAA tip as it was 20% inverted. Compressive, as well as some tensile, stresses were visible in the folded region of the LAA wall after simulation reached the 60% inversion. Only compressive stress was found in the LAA wall after the procedure was at 80% and 100% of the inversion. Local maxima of ~2.97 MPa compressive stresses were found at the end of the LAAI procedure.

Specifically, Figure 4A shows the stress profile as a function of the displacement of the LAA tip, and Figure 4B shows the pull force to invert the LAA. The stress at the end of simulation remains at the compressive state, with an average force to pull the LAA tip wall of 1.7 N for the whole simulation procedure.

Flow velocity at LAA was analyzed by streamlines at mitral valve flow peak of the E-wave and early diastole for both the non-inverted and LAAI models (Figure 5). At the peak velocity of E-wave, flow patterns were similar between the reference and LAAI models and were characterized by parallel flow streamlines with pronounced flow velocity at the branch of the pulmonary veins. At the early diastole, just after the deceleration phase of the A-wave, the flow velocity at the LAA was low with magnitude comparable to that of LA.

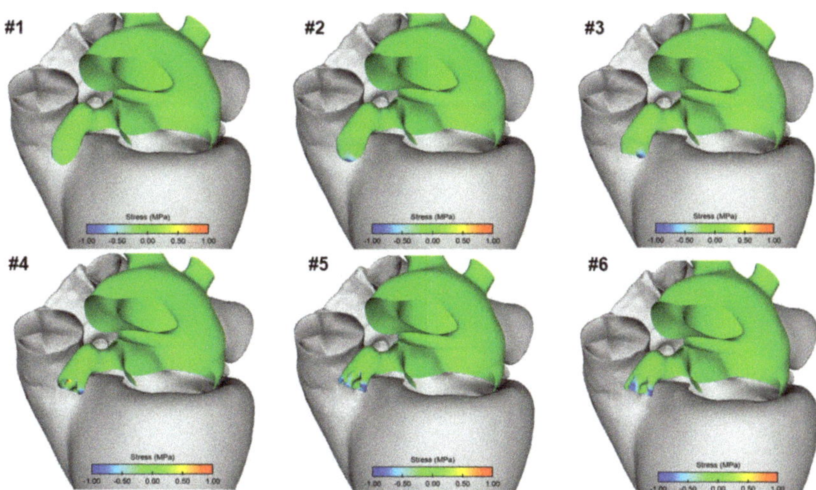

Figure 3. Color-coded stress distributions in the human left atrium during inversion of the LAA. Highest compressive (negative) stress (-1 MPa) is shown in dark blue; highest tensile (positive) stress (1 MPa) is shown in dark red. The progression of the procedure is from **#1** to **#6**. **#1**: Baseline, before any LAA inversion. **#2**: LAA is 20% inverted; compressive stress is seen near tip of LAA. **#3**: LAA is 40% inverted; compressive stress is seen over a larger portion of the LAA. **#4**: LAA is 60% inverted; compressive, as well as some tensile stress, is visible in LAA. **#5**: LAA is 80% inverted; only compressive stress is visible in LAA. **#6**: LAA is 100% inverted; only compressive stress is visible in LAA.

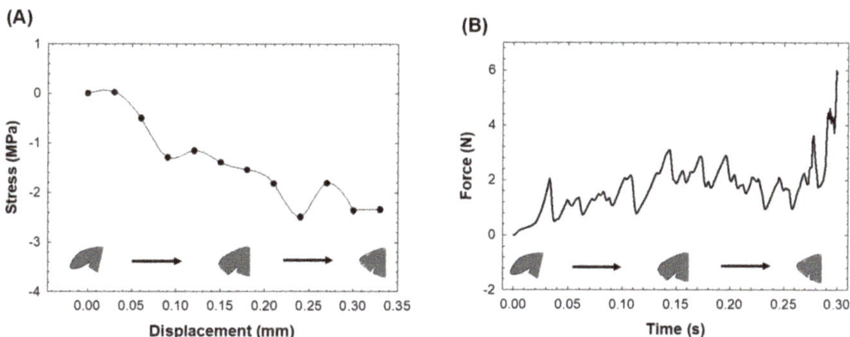

Figure 4. (**A**) Average stress at LAA tip as a function of displacement of the clamped nodes; (**B**) pulling force needed for LAAI as a function of simulation time.

From a qualitative perspective, there was a minimal difference in the flow pattern of the LAAI model versus the non-inverted reference model (Figure 6). Specifically, the LAAI model was characterized by a minimally-nested helical flow pattern with low velocity magnitude near the tip of the inverted appendage.

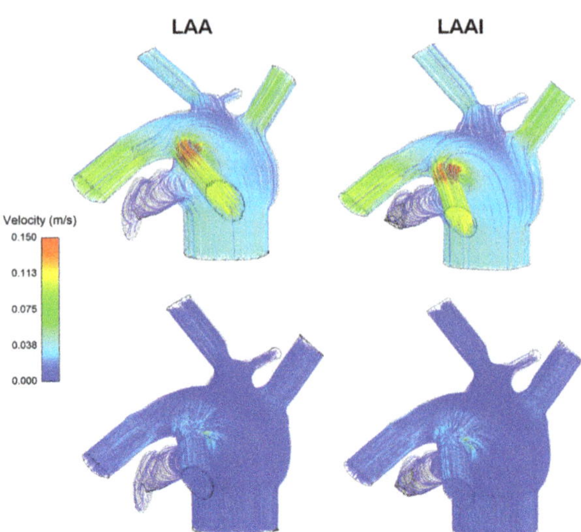

Figure 5. Flow velocity streamlines in the atrial chamber at peak systole of E-wave (**top** row) and diastole (**bottom** row) for both the non-inverted and inverted LAA.

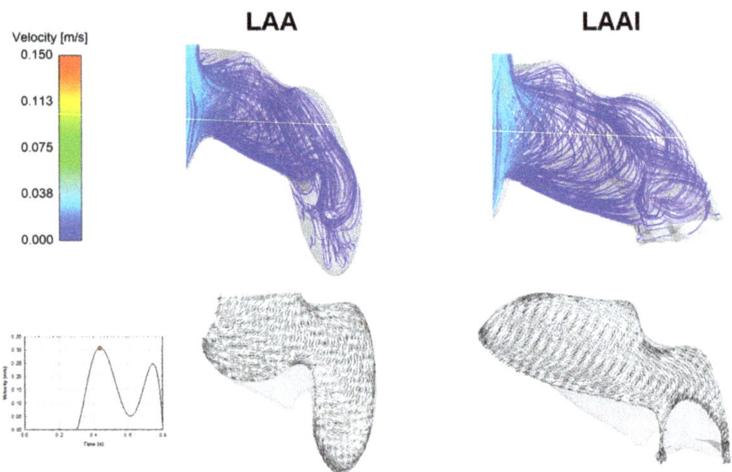

Figure 6. Flow velocity streamlines (top row) and velocity vector at a cross section for both the LAA and LAAI.

Figure 7 highlighted that the mean flow velocity in a cross section at the LAAI tip over one cardiac cycle was slightly higher than that computed for the non-inverted reference model during diastole and the E-wave.

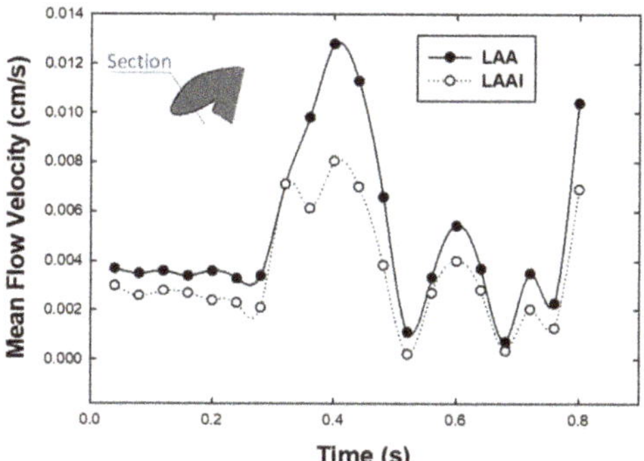

Figure 7. Mean flow velocity over one cardiac cycle at the tip of LAAI and in a cross-section of LAA (see inset).

4. Discussion

Using realistic left atrial anatomy as extracted from the LHHM, the LAA tip was virtually clamped, and the retraction procedure was simulated to reveal insights into the biomechanics of the inverted LAA tissue wall (Figures 1 and 2). The simulation confirmed the fundamental hypothesis of our study, suggesting a change in the stress distribution of the LAAI tissue wall from tensile to compressive (Figure 3) because of the inversion procedure. The circumferential stress of the LAAI tissue wall decreased upon 2.2 MPa at the end of inversion procedure, with the inversion process requiring a mean pull force of 1.7N (Figure 4). This compressive stress field can ultimately lead to resorption of the tissue as per reduced stress/resorption relation. Current LAA closure devices, whether epicardial or endocardial, must leave a device in the heart permanently, which may cause LAA perforation, migration, incomplete closure, new thrombus formation, and even thromboembolic events [26–28]. The concept of LAAI here proposed to promote resorption has not been previously reported and requires no permanent devices or implants (tissue glue or "stich" can be placed to ensure inversion configuration). We have here demonstrated the feasibility of such an approach by numerical simulation of LAAI in a realistic human model.

It has long been recognized that mechanics play a fundamental role in tissue growth and remodeling, especially in cardiac diseases. Mechanobiological control mechanisms in the myocardial wall tend to restore values of stress (or strain) toward preferred homeostatic values in response to diverse perturbations from normal conditions. For instance, hypertrophic growth of cardiomyocytes is the primary response by which the left ventricle reduces the stress on the myocardial wall imposed by pressure overload (e.g., hypertension) [29]. An increase in the tensile stress leads to intracellular signaling cascades that promote protein synthesis with consequent increases in the size and organization of cardiomyocytes and ultimately increasing the myocardium mass. Conversely, cardiac atrophy is a prevalent pathology associated with failed hearts after prolonged use of ventricular assist devices [30], resulting in an unloaded condition that causes tissue resorption. Based on these considerations, we expect that the compressive stress distribution generated on LAA wall by the inversion procedure could result in the resorption of the LAA tissue. Since the inverted state of the LAA offloads the stress (see Figure 3), we expect to resorb the apex ("blind end") of the LAA and hence eliminate the dead space. This hypothesis requires validation in chronic experiments because, although changes in atrial cardiomyocytes may occur within hours, changes at the cardiac wall level occur over days to weeks or months. The distribution of stress changes in the LAAI model can be considered as the basis for the development of mathematical models of cardiac growth and remodeling. Such an approach would be especially useful in understanding and predicting the long-term biological response of LAAI and its mechanistic link with the stress level exerted on the

LAA wall. The coupling of finite-element analysis with growth and remodeling, however, is complicated by the complexity of cardiac tissue.

From a hemodynamic perspective, the computational flow analysis revealed minimal differences in the flow patterns of the non-inverted LAA versus the LAAI (Figures 5 and 6), which was characterized by slightly reduced flow velocities. In this context, computational flow analysis has been used to predict the hemodynamic disturbances of LAA and the risk of complications in the setting of treated and untreated AF. Using moving wall capability, simulations predicted well the risk of LAA thrombosis in a small cohort of patients and demonstrated that both wall kinetics and LAA shape contribute the development of blood stagnation to the LAA [31]. Computational flow analysis also demonstrated that not only complex LAA shapes have low velocities and vorticity indexes and consequently high risk of thrombogenic events, but even simple morphologies may have thrombogenic risk equal to, or even higher than, more complex auricles [32]. Among LAA phenotypes, the Windsock LAA shape is associated with a high risk of thrombosis as compared to that of Cauliflower morphology, according to computational estimations of blood washout [33] and clinical evidence [34]. Our findings from computational flow analysis corroborate the low blood flow velocity pattern in the LAA geometry, and this suggests the risk of thrombosis for the LAAI model given the low velocity, which appeared similar to that of the non-inverted LAA. It is evident, however, that the LAAI procedure remarkably reduced the area of blood stasis because of inverted tissue wall occupying the auricle, ultimately determining AF-related ischemic stroke by detachment of thrombus material. Given the association between LAA phenotype and function, further studies on different LAA shapes are needed to better understand the development of thrombus formation after simulation of the LAAI procedure.

There are several limitations in this numerical proof of concept study. First, for the sake of simplicity, the LAA tissue wall was assumed to be a passive and isotropic material with uniform thickness. During heart beating, contractile material force is initiated through changes in the electrical potential and depends on reference tension, the primary fiber stress ratio, the fiber-stretch velocity and the current cellular state. Second, knowledge of myofiber orientation is a crucial for model development even in a non-contracting myocardium because myocardial mechanical properties are significantly stiffer in the local myofiber direction than in a plane transverse to the myofiber direction. Our group has extensively studied and developed constitutive material law to account for the active material contraction and includes myofiber orientation as validated against in-vivo measured strain data [35,36]. Further studies will be undertaken to consider a more realistic constitutive behavior for the atrial chamber to refine predictions of LAAI biomechanics and reduce the impact of model assumptions. Third, in computational flow analyses, the rigid LA wall did not allow us to include the LA volume variation induced by the atrial contraction, likely resulting in a flow-rate change over the cardiac cycle. Fourth, no turbulence model was included, even though recent studies demonstrated the presence of a transitional flow in specific regions of the atrial chamber, but depending on patient anatomy. Indeed, The LAA morphology can play an important role in the development of turbulent flow conditions. In fact, the cauliflower-type LAA may lead to low blood flow velocity and vorticity. In future studies, the impact of turbulence on the resulting hemodynamic of LAAI will be therefore investigated. However, this study was carried out to assess the overall impact of the LAAI hemodynamic, rather than perfecting the model representation and fidelity. Most importantly, this study was not developed to test the resorption of the inverted appendage for which a stress-growth law modeling the resorption of the biological tissue subjected to stress reduction is required. Finally, the present study was developed only on a single LAA geometry. Thus, the simulation framework here proposed will be applied in a large patient cohort to validate the results of the current proof of concept study.

5. Conclusions

As a proof-of-concept, this study demonstrated the feasibility of the LAAI procedure for removal of the dead space of the appendage without leaving any device behind. Although further model improvements are needed, the simulation framework here proposed can be used not only to quantify the biomechanics and hemodynamic of LAAI but also to optimize LAAI procedure development towards translation in clinical practice.

Author Contributions: Conceptualization, G.S.K. and J.M.G.; methodology, G.S.K., J.M.G. and S.P.; software, S.P.; formal analysis, G.S.K., J.M.G. and S.P.; writing—original draft preparation, G.S.K., J.M.G. and S.P.; writing—review and editing, G.S.K., J.M.G. and S.P.; funding acquisition, G.S.K. All authors have read and agreed to the published version of the manuscript.

Funding: This research was funded by NIH R43 HL149478-01.

Institutional Review Board Statement: Not applicable.

Informed Consent Statement: Not applicable.

Data Availability Statement: Not applicable.

Acknowledgments: The authors thank Jiang Yao of Dassault Systemes Simulia Corporation, for supporting the computational modeling and data analysis.

Conflicts of Interest: The funders had no role in the design of the study; in the collection, analyses, or interpretation of data; in the writing of the manuscript, or in the decision to publish the results.

References

1. Virani, S.S.; Alonso, A.; Aparicio, H.J.; Benjamin, E.J.; Bittencourt, M.S.; Callaway, C.W.; Carson, A.P.; Chamberlain, A.M.; Cheng, S.; Delling, F.N.; et al. Heart Disease and Stroke Statistics-2021 Update: A Report from the American Heart Association. *Circulation* **2021**, *143*, e254–e743. [CrossRef] [PubMed]
2. Patel, M.B.; Rasekh, A.; Shuraih, M.; Chelu, M.G.; Bartlett, T.; Mathuria, N.; Naeini, P.; Strickland, J.; Massumi, A.; Razavi, M.; et al. Safety and effectiveness of compassionate use of LARIAT(R) device for epicardial ligation of anatomically complex left atrial appendages. *J. Interv. Card. Electrophysiol.* **2015**, *42*, 11–19. [CrossRef] [PubMed]
3. January, C.T.; Wann, L.S.; Calkins, H.; Chen, L.Y.; Cigarroa, J.E.; Cleveland, J.C., Jr.; Ellinor, P.T.; Ezekowitz, M.D.; Field, M.E.; Furie, K.L.; et al. 2019 AHA/ACC/HRS Focused Update of the 2014 AHA/ACC/HRS Guideline for the Management of Patients with Atrial Fibrillation: A Report of the American College of Cardiology/American Heart Association Task Force on Clinical Practice Guidelines and the Heart Rhythm Society. *J. Am. Coll. Cardiol.* **2019**, *74*, 104–132. [CrossRef] [PubMed]
4. Kim, M.H.; Johnston, S.S.; Chu, B.C.; Dalal, M.R.; Schulman, K.L. Estimation of total incremental health care costs in patients with atrial fibrillation in the United States. *Circ. Cardiovasc. Qual. Outcomes* **2011**, *4*, 313–320. [CrossRef] [PubMed]
5. Camm, A.J.; Lip, G.Y.; De Caterina, R.; Savelieva, I.; Atar, D.; Hohnloser, S.H.; Hindricks, G.; Kirchhof, P.; Bax, J.J.; Baumgartner, H.; et al. 2012 focused update of the ESC Guidelines for the management of atrial fibrillation: An update of the 2010 ESC Guidelines for the management of atrial fibrillation. Developed with the special contribution of the European Heart Rhythm Association. *Eur. Heart J.* **2012**, *33*, 2719–2747. [CrossRef]
6. Al-Saady, N.M.; Obel, O.A.; Camm, A.J. Left atrial appendage: Structure, function, and role in thromboembolism. *Heart* **1999**, *82*, 547–554. [CrossRef]
7. Holmes, D.R., Jr.; Lakkireddy, D.R.; Whitlock, R.P.; Waksman, R.; Mack, M.J. Left atrial appendage occlusion: Opportunities and challenges. *J. Am. Coll. Cardiol.* **2014**, *63*, 291–298. [CrossRef]
8. Kakkar, A.K.; Mueller, I.; Bassand, J.P.; Fitzmaurice, D.A.; Goldhaber, S.Z.; Goto, S.; Haas, S.; Hacke, W.; Lip, G.Y.; Mantovani, L.G.; et al. Risk profiles and antithrombotic treatment of patients newly diagnosed with atrial fibrillation at risk of stroke: Perspectives from the international, observational, prospective GARFIELD registry. *PLoS ONE* **2013**, *8*, e63479. [CrossRef]
9. Molteni, M.; Cimminiello, C. Warfarin and atrial fibrillation: From ideal to real the warfarin affaire. *Thromb. J.* **2014**, *12*, 5. [CrossRef]
10. Holmes, D.R.; Reddy, V.Y.; Turi, Z.G.; Doshi, S.K.; Sievert, H.; Buchbinder, M.; Mullin, C.M.; Sick, P.; PROTECT AF Investigators. Percutaneous closure of the left atrial appendage versus warfarin therapy for prevention of stroke in patients with atrial fibrillation: A randomised non-inferiority trial. *Lancet* **2009**, *374*, 534–542. [CrossRef]
11. Cruz-Gonzalez, I.; Cubeddu, R.J.; Sanchez-Ledesma, M.; Cury, R.C.; Coggins, M.; Maree, A.O.; Palacios, I.F. Left atrial appendage exclusion using an Amplatzer device. *Int. J. Cardiol.* **2009**, *134*, e1–e3. [CrossRef] [PubMed]
12. Singh, S.M.; Dukkipati, S.R.; d'Avila, A.; Doshi, S.K.; Reddy, V.Y. Percutaneous left atrial appendage closure with an epicardial suture ligation approach: A prospective randomized pre-clinical feasibility study. *Heart Rhythm* **2010**, *7*, 370–376. [CrossRef] [PubMed]

13. Bakhtiary, F.; Kleine, P.; Martens, S.; Dzemali, O.; Dogan, S.; Keller, H.; Ackermann, H.; Zierer, A.; Ozaslan, F.; Wittlinger, T.; et al. Simplified technique for surgical ligation of the left atrial appendage in high-risk patients. *J. Thorac. Cardiovasc. Surg.* **2008**, *135*, 430–431. [CrossRef] [PubMed]

14. Salzberg, S.P.; Gillinov, A.M.; Anyanwu, A.; Castillo, J.; Filsoufi, F.; Adams, D.H. Surgical left atrial appendage occlusion: Evaluation of a novel device with magnetic resonance imaging. *Eur. J. Cardio-Thorac. Surg.* **2008**, *34*, 766–770. [CrossRef]

15. Turagam, M.K.; Vuddanda, V.; Verberkmoes, N.; Ohtsuka, T.; Akca, F.; Atkins, D.; Bommana, S.; Emmert, M.Y.; Gopinathannair, R.; Dunnington, G.; et al. Epicardial Left Atrial Appendage Exclusion Reduces Blood Pressure in Patients with Atrial Fibrillation and Hypertension. *J. Am. Coll. Cardiol.* **2018**, *72*, 1346–1353. [CrossRef]

16. Lakkireddy, D.; Turagam, M.; Afzal, M.R.; Rajasingh, J.; Atkins, D.; Dawn, B.; Di Biase, L.; Bartus, K.; Kar, S.; Natale, A.; et al. Left Atrial Appendage Closure and Systemic Homeostasis: The LAA HOMEOSTASIS Study. *J. Am. Coll. Cardiol.* **2018**, *71*, 135–144. [CrossRef]

17. Sulkin, M.S.; Berwick, Z.C.; Hermiller, J.B.; Navia, J.A.; Kassab, G.S. Suction catheter for enhanced control and accuracy of transseptal access. *EuroIntervention* **2016**, *12*, 1534–1541. [CrossRef] [PubMed]

18. Fung, Y.C. *Biomechanics: Motion, Flow, Stress and Growth*; Springer: Berlin/Heidelberg, Germany, 1990.

19. Peirlinck, M.; Costabal, F.S.; Yao, J.; Guccione, J.M.; Tripathy, S.; Wang, Y.; Ozturk, D.; Segars, P.; Morrison, T.M.; Levine, S.; et al. Precision medicine in human heart modeling: Perspectives, challenges, and opportunities. *Biomech. Model. Mechanobiol.* **2021**, *20*, 803–831. [CrossRef]

20. Holzapfel, G.A.; Ogden, R.W. Constitutive modelling of passive myocardium: A structurally based framework for material characterization. *Philos. Trans. A Math. Phys. Eng. Sci.* **2009**, *367*, 3445–3475. [CrossRef]

21. Rinaudo, A.; Pasta, S. Regional variation of wall shear stress in ascending thoracic aortic aneurysms. *Proc. Inst. Mech. Eng. Part H J. Eng. Med.* **2014**, *228*, 627–638. [CrossRef]

22. Pasta, S.; Gentile, G.; Raffa, G.M.; Scardulla, F.; Bellavia, D.; Luca, A.; Pilato, M.; Scardulla, C. Three-dimensional parametric modeling of bicuspid aortopathy and comparison with computational flow predictions. *Artif. Organs* **2017**, *41*, E92–E102. [CrossRef] [PubMed]

23. Rinaudo, A.; Raffa, G.M.; Scardulla, F.; Pilato, M.; Scardulla, C.; Pasta, S. Biomechanical implications of excessive endograft protrusion into the aortic arch after thoracic endovascular repair. *Comput. Biol. Med.* **2015**, *66*, 235–241. [CrossRef] [PubMed]

24. Mendez, V.; Di Giuseppe, M.; Pasta, S. Comparison of hemodynamic and structural indices of ascending thoracic aortic aneurysm as predicted by 2-way FSI, CFD rigid wall simulation and patient-specific displacement-based FEA. *Comput. Biol. Med.* **2018**, *100*, 221–229. [CrossRef] [PubMed]

25. Koizumi, R.; Funamoto, K.; Hayase, T.; Kanke, Y.; Shibata, M.; Shiraishi, Y.; Yambe, T. Numerical analysis of hemodynamic changes in the left atrium due to atrial fibrillation. *J. Biomech.* **2015**, *48*, 472–478. [CrossRef]

26. Lorenzoni, G.; Merella, P.; Pischedda, P.; Casu, G. Percutaneous Management of Left Atrial Appendage Perforation: Keep Calm and Think Fast. *J. Invasive Cardiol.* **2018**, *30*, E126–E127.

27. Katz, E.S.; Tsiamtsiouris, T.; Applebaum, R.M.; Schwartzbard, A.; Tunick, P.A.; Kronzon, I. Surgical left atrial appendage ligation is frequently incomplete: A transesophageal echocardiograhic study. *J. Am. Coll. Cardiol.* **2000**, *36*, 468–471. [CrossRef]

28. Donnino, R.; Tunick, P.A.; Kronzon, I. Left atrial appendage thrombus outside of a 'successful' ligation. *Eur. J. Echocardiogr.* **2008**, *9*, 397–398. [CrossRef]

29. Grossman, W. Cardiac hypertrophy: Useful adaptation or pathologic process? *Am. J. Med.* **1980**, *69*, 576–584. [CrossRef]

30. Diakos, N.A.; Selzman, C.H.; Sachse, F.B.; Stehlik, J.; Kfoury, A.G.; Wever-Pinzon, O.; Catino, A.; Alharethi, R.; Reid, B.B.; Miller, D.V.; et al. Myocardial atrophy and chronic mechanical unloading of the failing human heart: Implications for cardiac assist device-induced myocardial recovery. *J. Am. Coll. Cardiol.* **2014**, *64*, 1602–1612. [CrossRef]

31. Garcia-Villalba, M.; Rossini, L.; Gonzalo, A.; Vigneault, D.; Martinez-Legazpi, P.; Duran, E.; Flores, O.; Bermejo, J.; McVeigh, E.; Kahn, A.M.; et al. Demonstration of Patient-Specific Simulations to Assess Left Atrial Appendage Thrombogenesis Risk. *Front. Physiol.* **2021**, *12*, 596596. [CrossRef]

32. Masci, A.; Barone, L.; Dede, L.; Fedele, M.; Tomasi, C.; Quarteroni, A.; Corsi, C. The Impact of Left Atrium Appendage Morphology on Stroke Risk Assessment in Atrial Fibrillation: A Computational Fluid Dynamics Study. *Front. Physiol.* **2018**, *9*, 1938. [CrossRef] [PubMed]

33. Bosi, G.M.; Cook, A.; Rai, R.; Menezes, L.J.; Schievano, S.; Torii, R.; Burriesci, G.B. Computational Fluid Dynamic Analysis of the Left Atrial Appendage to Predict Thrombosis Risk. *Front. Cardiovasc. Med.* **2018**, *5*, 34. [CrossRef] [PubMed]

34. Di Biase, L.; Santangeli, P.; Anselmino, M.; Mohanty, P.; Salvetti, I.; Gili, S.; Horton, R.; Sanchez, J.E.; Bai, R.; Mohanty, S.; et al. Does the left atrial appendage morphology correlate with the risk of stroke in patients with atrial fibrillation? Results from a multicenter study. *J. Am. Coll. Cardiol.* **2012**, *60*, 531–538. [CrossRef] [PubMed]

35. Sack, K.L.; Aliotta, E.; Ennis, D.B.; Choy, J.S.; Kassab, G.S.; Guccione, J.M.; Franz, T. Construction and Validation of Subject-Specific Biventricular Finite-Element Models of Healthy and Failing Swine Hearts from High-Resolution DT-MRI. *Front. Physiol.* **2018**, *9*, 539. [CrossRef] [PubMed]

36. Sack, K.L.; Aliotta, E.; Choy, J.S.; Ennis, D.B.; Davies, N.H.; Franz, T.; Kassab, G.S.; Guccione, J.M. Intra-myocardial alginate hydrogel injection acts as a left ventricular mid-wall constraint in swine. *Acta Biomater.* **2020**, *111*, 170–180. [CrossRef] [PubMed]

Journal of
Personalized Medicine

Article

Clinical Phenotypes of Atrial Fibrillation and Mortality Risk—A Cluster Analysis from the Nationwide Italian START Registry

Daniele Pastori [1,*], Emilia Antonucci [2], Alberto Milanese [3], Danilo Menichelli [1], Gualtiero Palareti [2], Alessio Farcomeni [4], Pasquale Pignatelli [1] and on behalf of START2 Register Investigators [†]

1. Department of Clinical, Internal, Anesthesiological, and Cardiovascular Sciences, Sapienza University of Rome, 00185 Rome, Italy; danilo.menichelli@uniroma1.it (D.M.); pasquale.pignatelli@uniroma1.it (P.P.)
2. Arianna Anticoagulazione Foundation, 40138 Bologna, Italy; e.antonucci@fondazionearianna.org (E.A.); gualtiero.palareti@unibo.it (G.P.)
3. Department of Public Health and Infectious Diseases, Sapienza University of Rome, 00185 Rome, Italy; alberto.milanese@uniroma1.it
4. Department of Economics and Finance, University of Rome "Tor Vergata", 00133 Rome, Italy; alessio.farcomeni@uniroma2.it
* Correspondence: daniele.pastori@uniroma1.it; Tel.: +39-06-4997-2300; Fax: +39-06-4997-2309
† Members of the START2 Register Investigators is provided in Appendix A.

Abstract: Patients with atrial fibrillation (AF) still experience a high mortality rate despite optimal antithrombotic treatment. We aimed to identify clinical phenotypes of patients to stratify mortality risk in AF. Cluster analysis was performed on 5171 AF patients from the nationwide START registry. The risk of all-cause mortality in each cluster was analyzed. We identified four clusters. *Cluster 1* was composed of the youngest patients, with low comorbidities; *Cluster 2* of patients with low cardiovascular risk factors and high prevalence of cancer; *Cluster 3* of men with diabetes and coronary disease and peripheral artery disease; *Cluster 4* included the oldest patients, mainly women, with previous cerebrovascular events. During 9857 person-years of observation, 386 deaths (3.92%/year) occurred. Mortality rates increased across clusters: 0.42%/year (cluster 1, reference group), 2.12%/year (cluster 2, adjusted hazard ratio (aHR) 3.306, 95% confidence interval (CI) 1.204–9.077, $p = 0.020$), 4.41%/year (cluster 3, aHR 6.702, 95%CI 2.433–18.461, $p < 0.001$), and 8.71%/year (cluster 4, aHR 8.927, 95%CI 3.238–24.605, $p < 0.001$). We identified four clusters of AF patients with progressive mortality risk. The use of clinical phenotypes may help identify patients at a higher risk of mortality.

Keywords: atrial fibrillation; all-cause mortality; phenotype; risk factors

Citation: Pastori, D.; Antonucci, E.; Milanese, A.; Menichelli, D.; Palareti, G.; Farcomeni, A.; Pignatelli, P.; on behalf of START2 Register Investigators. Clinical Phenotypes of Atrial Fibrillation and Mortality Risk—A Cluster Analysis from the Nationwide Italian START Registry. *J. Pers. Med.* **2022**, *12*, 785. https://doi.org/10.3390/jpm12050785

Academic Editor: Oscar Campuzano

Received: 20 March 2022
Accepted: 10 May 2022
Published: 12 May 2022

1. Introduction

Atrial fibrillation (AF) is a highly prevalent cardiac disease characterized by an increased risk of thromboembolic events and cardiovascular disease, such as myocardial infarction [1]. In addition to ischemic complications, patients with AF experience a high rate of mortality, which is estimated at $\geq 4\%$/year [2,3]. Of note, at least one-third of causes of death are related to non-cardiovascular disease [4,5], which are indeed not significantly modified by antithrombotic treatments.

Despite the extensive use of scores for risk stratification in AF, this approach presents several limitations.

Thus, a recent study confirmed that in AF patients the predictive performance of common risk scores against mortality was limited, with c-indexes generally <0.65 [6]. Some other scores are also difficult to calculate, requiring many clinical and laboratory variables and are therefore not easy to be used in daily clinical practice [7,8]. Furthermore, in the need of a simple approach, some important clinical characteristics are often not included in current risk stratification schemes, neglecting some potential important factors that need to be addressed. In many cases, scores are also applied to cohorts with missing variables or with single items not properly calculated as in the case of retrospective registries using codes. Of note, in patients with AF, only one risk score, the BASIC-AF risk score has been proposed to predict mortality in AF patients [9]. However, this score includes imaging and laboratory variables that are not always available for outpatients [9].

Cluster analysis may play a role in overcoming these limitations, especially in the case of overlapping risk factors. Previous studies showed that clustering may allow a better characterization of the disease phenotype in different clinical settings such as heart failure [10,11] and chronic obstructive pulmonary disease [12]. This approach may have an important impact in clinical practices by implementing risk stratification.

The aim of the study was to analyze in patients enrolled in the large cohort of the START (Survey on anTicoagulated pAtients RegisTer) registry, clinical phenotypes of AF by cluster analysis, and the association with mortality risk.

2. Materials and Methods

Details of the multicenter nationwide START registry were previously described [13]. Briefly, the START-register is an observational, multicenter, ongoing cohort study that includes patients ($\geq$18 years) who start anticoagulation therapy. The present analysis is limited to patients with non-valvular AF starting oral anticoagulants, either vitamin K antagonists (VKAs) or direct oral anticoagulants (DOACs). Patients treated with low-molecular weight heparin were excluded. Patients with life expectancy <6 months, or non-residents in the participant region, or planning to leave in the next 6 months, were not included in the registry, as well as patients already enrolled in phase II or III clinical studies. Patients enrolled in other observational or phase IV studies were considered eligible for the study.

2.1. Ethics

All patients signed an informed written consent at study entry. The registry was approved in October 2011 (ref. 142/2010/0/0ss) by the Ethical Committee of the Institution of the Coordinating Member (University Hospital "S. Orsola-Malpighi", Bologna, Italy). The study is registered at clinicaltrials.gov identifier: NCT02219984 and is still ongoing/recruitment is still open. The study is conducted according to the declaration of Helsinki.

In particular, the START registry (Survey on anTicoagulated pAtients RegisTer, NCT02219984), is promoted by the Arianna Anticoagulazione Foundation, Bologna. The registry is investigator-driven, non-sponsored, and was approved by the ethics committee of each participating institution (Campus Bio-Medico University of Rome, Italy; Monaldi Hospital and "Luigi Vanvitelli" University of Campania, Italy; "Federico II" University of Naples, Italy; University of Perugia, Italy; University Hospital of Padua, Italy; Sapienza University of Rome, Italy; University of Florence, Italy).

2.2. Patient and Public Involvement Statement

Patients or the public were not involved in the design, conduct, reporting, or dissemination plans of our research.

2.3. Statistical Analysis

Data are expressed as mean and standard deviation or the median and interquartile range (IQR) depending on the variable distribution. Group comparisons were performed by the unpaired Student's t-test. Proportions and categorical variables were tested by the χ^2 test.

In order to identify subgroups of patients with the most similar baseline characteristics we selected a pool of variables and proceeded with cluster analysis. We decided to use the following clinical variables: age, sex, diabetes, previous cerebrovascular events (defined as ischemic stroke or transient ischemic attack), previous cardiovascular events, heart failure, peripheral artery disease (PAD), use of non-vitamin K oral anticoagulants, cancer, pulmonary disease, smoking habit, previous major bleeding. The following clinical variables were instead not used for the cluster analysis: persistent/permanent AF, body mass index (BMI), hypertension. We excluded these variables as their use would have led to a large number of groups, which would not have been useful for clinical purposes. Concomitant drugs were not used for clustering to avoid bias by indication. Furthermore, composite variables, such as CHA_2DS_2-VASc and HAS-BLED scores were not used since we used their components. For cluster analysis we proceeded using a model based procedure where continuous variables in each cluster were assumed to follow the multivariate Gaussian distribution, and categorical variables to follow a multinomial distribution, as in Hennig and Liao [14]. Model-based clustering allows us to use a formal criterion for selecting the optimal number of groups. In this work, we selected the optimal number of clusters by comparing minimizing the Bayesian information criterion.

The incidence of all-cause mortality by each cluster was estimated using a Kaplan–Meier product-limit estimator. Survival curves were formally compared using the log-rank test. Univariable and multivariable Cox proportional hazard regression analyses were used to calculate the (adjusted) relative hazard ratios (HRs) of death. In the multivariable model, we adjusted for variables not used to define the clusters, not using composite variables (such as risk scores) to avoid overadjustment.

All tests were two-tailed, a p-value < 0.05 was considered statistically significant. Analyses were performed using computer software IBM® SPSS® Statistics version 25 (IBM, Armonk, NY, USA). and R version 3.6.2 (RStudio, Boston, MA, USA).

3. Results

3.1. Description of Clusters

Table 1 shows clinical and biochemical characteristics of patients according to each cluster.

Cluster 1 (n = 512). Youngest and low comorbidities. This cluster included patients with the lowest mean age (55.6 ± 7.9 years) and with a low prevalence of women (23.6%), and with the overall lowest burden of cardiovascular comorbidities with only 14.6% of patients with a history of cerebrovascular events (second lowest prevalence). These patients were less likely to be treated with DOACs (only 10%). In this cluster, there was the highest proportion of obese patients (30.1%) and the highest use of anti-arrhythmic drugs (32.8%), probably related to the low proportion of patients with persistent/permanent AF (51.4%).

Cluster 2 (n = 2201). Low cardiovascular risk and high cancer. This was the largest cluster including patients with a relatively high mean age (75.0 ± 6.0 years) and 54% of patients were women (second highest group). This group was characterized by the lowest prevalence of cardiovascular risk factors and the highest proportion of patients with cancer (18.2%). Regarding anticoagulant treatment, the use of DOACs was present in 27% of patients. Among medications, aspirin was prescribed in 5.2% of patients despite a very low prevalence of cardiovascular disease at baseline (1.5%). The prevalence of anemia was significantly higher than cluster 1 (19.3% versus 11.3%, respectively).

Cluster 3 (n = 1268). High cardiovascular risk and more men. This cluster displayed the lowest prevalence of women (8.1%) while the mean age was similar to cluster 2. Cardiovascular risk factors were highly prevalent, being the cluster with the highest proportion of diabetes (35.0%), previous cardiovascular disease (53.5%), PAD (16.1%), and chronic pulmonary disease (27.8%). Moreover, previous cerebrovascular events (17.2%) and heart failure (29.2%) were common, with the second highest prevalence among clusters. This cluster disclosed the highest use of aspirin in 21.1%. Among variables not used for clustering, this group was the first for highest for the use of proton pump inhibitors (58.7%) and statins (54.7%) and for the prevalence of thrombocytopenia (14.7%), and the second highest for the prevalence of anemia and hypertension.

Cluster 4 (n = 1190). Oldest, more women, and cerebrovascular disease. This cluster was composed mainly by elderly patients (mean age 83.7 ± 4.2) with the highest number of women (78.2%) and persistent/permanent AF (70.8%). The prevalence of previous cerebrovascular events in this group was the highest among clusters with 23.9% of patients, as well as heart failure (31.1%); the second highest proportion of chronic pulmonary disease (21.0%) and previous cardiovascular disease (18.1%) was found. This was the group with the highest use of DOACs (33.8%). Regarding other comorbidities, this group had the highest prevalence of chronic kidney disease (78.8%), anemia (34.6%), and hypertension (88.2%). Concerning medications, this cluster had the highest use of DOACs (33.8%), digoxin (15.8%), and the second highest use of statins (29.5%) and proton pump inhibitors (53.1%).

Table 1. Clinical and biochemical characteristics of patients according to each cluster.

Cluster Denomination	Whole Cohort	Cluster 1	Cluster 2	Cluster 3	Cluster 4	*p*-Value (among Groups)
		Youngest and Low Comorbidities	Low Cardiovascular Risk and High Cancer	High Cardiovascular Risk and More Men	Oldest, More Women and Cerebrovascular Disease	
Cluster size *n*	5171	512	2201	1268	1190	
			Variables used to define clusters			
Age (years)	75.0 ± 9.6	55.6 ± 7.9	75.0 ± 6.0	74.6 ± 7.0	83.7 ± 4.2	<0.001
Women (%)	45.3	23.6	54.0	8.1	78.2	<0.001
Diabetes (%)	20.2	10.7	16.5	35.0	15.1	<0.001
Previous cerebrovascular events (%)	16.5	14.6	12.5	17.2	23.9	<0.001
Previous cardiovascular disease (%)	18.6	6.8	1.5	53.5	18.1	<0.001
Heart failure (%)	15.5	7.0	1.1	29.2	31.1	<0.001
Peripheral Artery Disease (%)	6.4	0.8	0.6	16.1	9.1	<0.001
Cancer (%)	13.6	2.9	18.2	15.1	8.1	<0.001
Pulmonary disease (%)	12.6	3.1	1.5	27.8	21.0	<0.001
Smoking (%)	13.2	21.9	2.7	39.4	1.1	<0.001
Previous major bleeding (%)	3.5	1.4	1.9	4.5	6.1	<0.001
DOACs (vs. VKAs) (%)	25.8	10.0	27.0	22.7	33.8	<0.001
			Variables not used for cluster analysis			
Persistent/permanent AF (%)	63.3	51.4	60.7	65.7	70.8	<0.001
BMI (kg/m^2)	26.9 ± 4.7	28.1 ± 5.5	26.7 ± 4.5	27.6 ± 4.6	25.8 ± 4.6	<0.001
Obesity (BMI ≥ 30 kg/m^2)	21.1	30.1	19.9	24.1	16.6	<0.001
Creatinine Clearance (mL/min)	66.8 ± 28.3	103.8 ± 33.6	67.6 ± 22.8	68.6 ± 26.8	47.6 ± 17.4	<0.001
Chronic kidney disease (Creatinine clearance <60 mL/min) (%)	45.1	5.1	39.5	39.4	78.8	<0.001
Hemoglobin (g/dl)	13.5 ± 1.8	14.5 ± 1.6	13.6 ± 1.6	13.6 ± 1.8	12.7 ± 1.6	<0.001
Anemia (<12 g/dL for women and <13 g/dL for men) (%)	24.7	11.3	19.3	30.0	34.6	<0.001
Platelet count ($\times 10^9$/L)	222.2 ± 68.9	223.3 ± 62.0	223.0 ± 69.6	213.8 ± 70.7	229.2 ± 67.7	<0.001
Thrombocytopenia (<150 $\times 10^9$/L, %)	10.7	9.0	10.3	14.7	7.9	<0.001
Hypertension (%)	80.6	59.6	78.1	86.3	88.2	<0.001
CHA$_2$DS$_2$ VASc score	3.6 ± 1.5	1.5 ± 1.1	3.3 ± 1.2	3.9 ± 1.4	4.7 ± 1.2	<0.001
HAS-BLED score	1.3 ± 0.7	0.4 ± 0.6	1.2 ± 0.6	1.5 ± 0.8	1.5 ± 0.6	<0.001
Aspirin (%)	9.7	6.3	5.2	21.1	7.5	<0.001
Statins (%)	33.7	21.3	26.8	54.7	29.5	<0.001
Anti-arrhythmic drugs (%)	25.2	32.8	26.2	25.1	20.3	<0.001
Digoxin (%)	9.2	6.1	7.2	8.0	15.8	<0.001
Proton pump inhibitors (%)	45.9	32.6	37.8	58.7	53.1	<0.001

BMI: body mass index; DOAC: direct oral anticoagulant; VKA: vitamin K antagonist; AF: atrial fibrillation; BMI, body mass index.

3.2. Clusters and Mortality Risk

During a mean follow-up of 22.9 ± 16.7 months yielding 9857 person-years of observation, 386 deaths (3.92%/year) were registered. Incidence rate of mortality was 0.42%/year (95%CI 0.11–1.10) in cluster 1 (reference group), 2.12%/year (95%CI 1.71–2.59) in cluster 2 (HR 5.068, 95%CI 1.863–13.784, $p = 0.001$ versus cluster 1), 4.41%/year (95%CI 3.60–5.35) in cluster 3 (HR 10.513, 95%CI 3.872–28.544, $p < 0.001$ versus cluster 1), 8.71%/year (95%CI 7.50–10.1) in cluster 4 (HR 20.708, 95%CI 7.690–55.761, $p < 0.001$ versus cluster 1) (Figure 1).

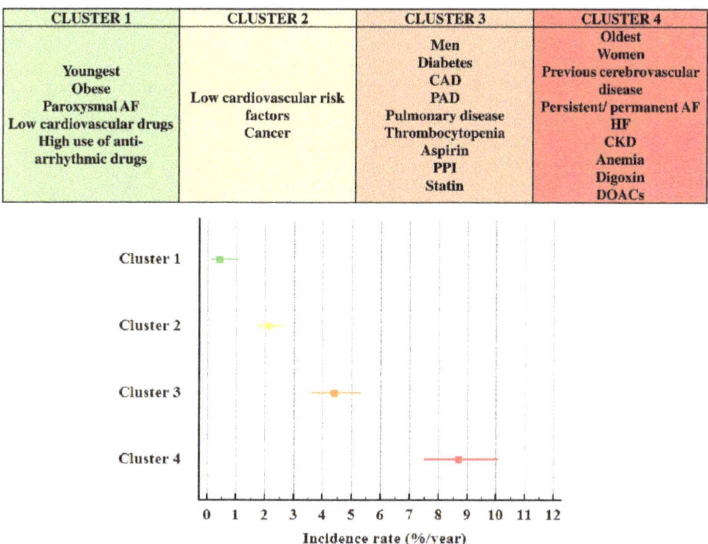

Figure 1. Description of clusters characteristics and incidence rates of mortality. Abbreviations: AF: atrial fibrillation; CAD: coronary artery disease; PAD: peripheral artery disease; PPI: proton pump inhibitors; HF: heart failure; CKD: chronic kidney disease; DOACs: direct oral anticoagulants.

Kaplan–Meier curves (Figure 2) showed a significant difference in the incidence of mortality across clusters, which increased from cluster 1 to 4 (log-rank test $p < 0.001$).

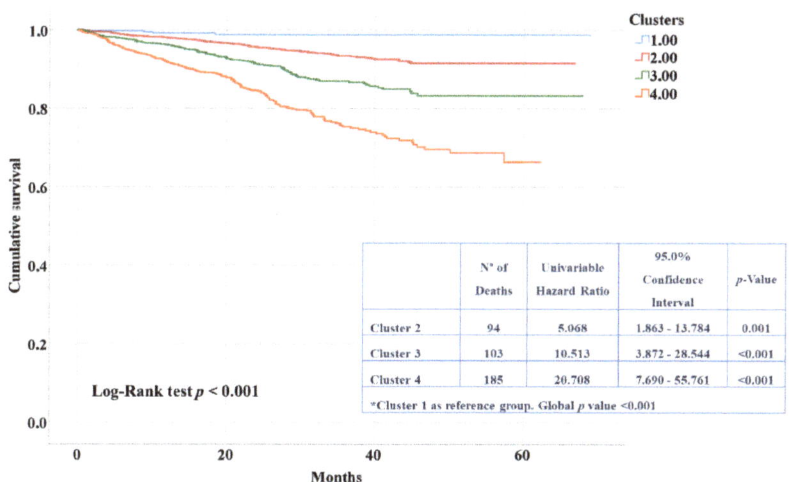

	N° of Deaths	Univariable Hazard Ratio	95.0% Confidence Interval	*p*-Value
Cluster 2	94	5.068	1.863 - 13.784	0.001
Cluster 3	103	10.513	3.872 - 28.544	<0.001
Cluster 4	185	20.708	7.690 - 55.761	<0.001
*Cluster 1 as reference group. Global *p* value <0.001				

Figure 2. Kaplan–Meier curves for risk of mortality according to different clusters.

At the multivariable Cox proportional regression analysis, clusters remained associated with mortality after adjustment for confounding factors and medications (Table 2).

Table 2. Multivariable Cox proportional regression analysis of factors associated with mortality.

Variables	Hazard Ratio	95% Confidence Interval		*p*-Value
Cluster 2 (vs. 1) *	**3.306**	**1.204**	**9.077**	**0.020**
Cluster 3 (vs. 1) *	**6.702**	**2.433**	**18.461**	**<0.001**
Cluster 4 (vs. 1) *	**8.927**	**3.238**	**24.605**	**<0.001**
Persistent/permanent AF	1.231	0.975	1.553	0.081
Statin	**0.655**	**0.519**	**0.828**	**<0.001**
Digoxin	0.963	0.692	1.339	0.822
Proton pump inhibitors	**1.367**	**1.108**	**1.686**	**0.004**
Hypertension	1.009	0.747	1.363	0.953
Obesity	1.217	0.930	1.592	0.152
Anemia	**1.618**	**1.313**	**1.993**	**<0.001**
Thrombocytopenia	**1.418**	**1.060**	**1.898**	**0.019**
Chronic kidney disease	**2.347**	**1.821**	**3.024**	**<0.001**
Anti-arrhythmic drugs	**0.713**	**0.552**	**0.922**	**0.010**
Aspirin	0.880	0.620	1.248	0.472

* Global *p*-value *p* < 0.001. Abbreviation: AF: atrial fibrillation. Statistically significant values are marked with bold

4. Discussion

From the large dataset of the Italian START registry, we identified four groups of AF patients with specific characteristics and graded progressive risk of all-cause mortality.

The four clusters showed specific clinical characteristics. Cluster 1 was the group with the lowest incidence of mortality and was composed of the youngest patients, with obesity and low comorbidities. This group was coincidentally characterized by a relatively lower proportion of paroxysmal AF, compared to the other clusters. The lower risk of mortality in patients with paroxysmal AF was reported in the post hoc analysis of the ENGAGE AF-TIMI 48 Trial (Effective Anticoagulation with Factor Xa Next Generation in Atrial Fibrillation-Thrombolysis in Myocardial Infarction 48), which showed a lower risk of mortality in paroxysmal versus permanent AF (1.49%/year and 1.95%/year, respectively) [15]. This cluster was also characterized by a higher proportion of obese patients. Regarding obesity, the real meaning of this association is hard to explain, as BMI does not take into account fat composition and visceral adiposity. A recent analysis showed a U-shaped association between body weight and mortality in AF [16].

Cluster 2 included patients with low cardiovascular risk factors and a high proportion of cancer. Patients in this group disclosed a five-fold increased risk of mortality compared to patients without cancer. This finding is in line with a previous finding showing that cancer is an important risk factor for mortality in the AF population [5,17], requiring specific management of anticoagulation according to cancer-specific treatments [18].

Cluster 3 was composed of mainly men with diabetes and coronary and PAD, a high proportion of thrombocytopenia, and a high use of aspirin, proton pump inhibitors, and statins. This cluster clearly defines patients with vascular disease. Coronary disease and PAD are frequently associated in patients with AF and increase the risk of cardiovascular events [19]. Thus, it is not surprising that in this cluster there was a high use of statins, which are recommended to prevent cardiovascular events in patients with PAD [20] and have been shown to improve outcomes in the AF population [21].

Cluster 4 included the oldest patients, mainly women, with previous cerebrovascular events, persistent/permanent AF, heart failure, kidney disease and anemia. The lowest prevalence of obesity in this cluster may reflect the association of sarcopenia with advancing age. The high proportion of patients with heart failure in this cluster (>30%) confirms the pivotal role of this comorbidity as the leading cause of death in patients with AF, even more important than ischemic stroke [22,23].

Our analysis has significant differences when compared to two previous studies that analyzed the clinical phenotype of AF patients [24,25]. The first study developed clusters from a Japanese cohort with very different clinical characteristics than our cohort, such as low prevalence of hypertension, which is actually one of the leading causes of mortality worldwide [24]. Furthermore, about 11% of patients with an indication to oral anticoagulation, i.e., a CHA_2DS_2-VASc score equal to or above 2, were not taking anticoagulants [26]. Another study included 9749 patients with AF in the US [25], and identified four clusters which, however, were not easy to use; thus, cluster 3 shared similar characteristics of cluster 4 regarding the proportion of hypertension, respiratory and chronic kidney disease, along with a similar age [25]. Furthermore, the proportion of diabetes was very similar between clusters 3 and 2 [25]. This overlap of risk factors was also evident in the external cohort from the ORBIT-AF [25]. All these factors make the allocation of patients in a specific cluster difficult. Finally, cancer was not considered for cluster formation in either of the two studies; this is an important point as we believe that it may define a specific subgroup of AF patients with peculiar characteristics.

The identification of clinical clusters of AF patients at different mortality risks may be complementary to the integrated approach proposed by recent guidelines for the management of AF patients [26,27]. In this view, the application of the ABC pathway may differ in the four clusters. For instance, patients in cluster 1 may benefit from an early rhythm and symptoms control, whilst patients in clusters 3 and 4 from a tight control of cardiometabolic diseases, such as diabetes, hypertension, and dyslipidemia. This tailored approach may lead to a reduction of mortality in this population of high-risk patients.

Our study has some strengths and limitations to acknowledge. Cluster analysis is an unbiased approach to identify patients at a higher risk of death. Thus, apart from oral anticoagulation, which is a common indication for all patients, we did not use concomitant drugs to define clusters, to avoid any bias by indication. Moreover, we also considered risk factors not included in the current risk stratification scores, providing information on additional comorbidities that need to be managed in AF patients. The large sample size of the study cohort, which recruited patients from any region of our country, is another strength of our work, which makes our study representative of our general AF population and adequate to perform the cluster analysis. Despite the advantage provided by a cluster-based approach to identify subgroups of patients within a specific disease, the problem with the cluster analysis is the generalization of results to other populations [28].

As a limitation, and an open field for future research, we could not investigate the association of clusters with specific causes of death. We only included patients of Caucasian ethnicity; thus, clinical phenotypes may be different in other populations.

In conclusion, we identified specific phenotypes of AF patients showing a different association with mortality. A correct global risk stratification strategy should include clinical phenotypes of patients beyond risk scores application.

Author Contributions: Conceptualization, D.P.; writing—original draft preparation, D.P., D.M., P.P.; data curation, E.A. and G.P.; formal analysis, D.P., A.M. and A.F.; statistical analysis, A.M. and A.F. All authors have read and agreed to the published version of the manuscript.

Funding: Publication fees for this article have been covered by Progetto di Ateneo 2022 assigned to Daniele Pastori (ref. RM12117A82C9EE21) by Sapienza University of Rome.

Institutional Review Board Statement: The registry was approved in October 2011 (n = 142/2010/0/0ss) by the Ethics Committee of the Institution of the Coordinating Member (University Hospital "S. Orsola-Malpighi", Bologna, Italy). The study is registered at clinicaltrials.gov identifier: NCT02219984 and is still ongoing/recruitment is still open. The study was conducted in accordance with the Declaration of Helsinki.

Informed Consent Statement: Informed consent was obtained from all subjects involved in the study.

Data Availability Statement: The data underlying this article will be shared upon reasonable request to the corresponding author.

Acknowledgments: This manuscript was accepted as the best poster presentation at the European Society of Cardiology (ESC) Congress 2020, and at the 121st Congress of the Italian Society of Internal Medicine (SIMI).

Conflicts of Interest: The authors declare no conflict of interest.

Appendix A

The START2 Register Investigators: Sophie Testa, Oriana Paoletti, Benilde Cosmi, Giuliana Guazzaloca, Ludovica Migliaccio, Daniela Poli, Rossella Marcucci, Niccolò Maggini, Vittorio Pengo, Anna Falanga, Teresa Lerede, Lucia Ruocco, Giuliana Martini, Simona Pedrini, Federica Bertola, Lucilla Masciocco, Pasquale Saracino, Angelo Benvenuto, Claudio Vasselli, Pasquale Pignatelli, Daniele Pastori, Danilo Menichelli Elvira Grandone, Donatella Colaizzo, Marco Marzolo, Mauro Pinelli, Daniela Mastroiacovo, Walter Ageno, Giovanna Colombo, Eugenio Bucherini, Domizio Serra, Andrea Toma, Pietro Barbera, Carmelo Paparo, Antonio Insana, Serena Rupoli, Giuseppe Malcangi, Maddalena Loredana Zighetti, Catello Mangione, Domenico Lione, Paola Casasco, Giovanni Nante, Alberto Tosetto, Vincenzo Oriana, Nicola Lucio Liberato.

References

1. Pastori, D.; Menichelli, D.; Del Sole, F.; Pignatelli, P.; Violi, F.; ATHERO-AF study group. Long-Term Risk of Major Adverse Cardiac Events in Atrial Fibrillation Patients on Direct Oral Anticoagulants. *Mayo Clin. Proc.* **2020**, *96*, 658–665. [CrossRef] [PubMed]
2. Pastori, D.; Antonucci, E.; Violi, F.; Palareti, G.; Pignatelli, P. Thrombocytopenia and Mortality Risk in Patients With Atrial Fibrillation: An Analysis From the START Registry. *J. Am. Heart Assoc.* **2019**, *8*, e012596. [CrossRef] [PubMed]
3. Perera, K.S.; Pearce, L.A.; Sharma, M.; Benavente, O.; Connolly, S.J.; Hart, R.G.; ACTIVE A (Atrial Fibrillation Clopidogrel Trial With Irbesartan for Prevention of Vascular Events) Steering Committee and Investigators. Predictors of Mortality in Patients With Atrial Fibrillation (from the Atrial Fibrillation Clopidogrel Trial With Irbesartan for Prevention of Vascular Events [ACTIVE A]). *Am. J. Cardiol.* **2018**, *121*, 584–589. [CrossRef] [PubMed]
4. Fauchier, L.; Villejoubert, O.; Clementy, N.; Bernard, A.; Pierre, B.; Angoulvant, D.; Ivanes, F.; Babuty, D.; Lip, G.Y. Causes of Death and Influencing Factors in Patients with Atrial Fibrillation. *Am. J. Med.* **2016**, *129*, 1278–1287. [CrossRef]
5. Gomez-Outes, A.; Lagunar-Ruiz, J.; Terleira-Fernandez, A.I.; Calvo-Rojas, G.; Suarez-Gea, M.L.; Vargas-Castrillon, E. Causes of Death in Anticoagulated Patients With Atrial Fibrillation. *J. Am. Coll. Cardiol.* **2016**, *68*, 2508–2521. [CrossRef] [PubMed]
6. Proietti, M.; Farcomeni, A.; Romiti, G.F.; Di Rocco, A.; Placentino, F.; Diemberger, I.; Lip, G.Y.; Boriani, G. Association between clinical risk scores and mortality in atrial fibrillation: Systematic review and network meta-regression of 669,000 patients. *Eur. J. Prev. Cardiol.* **2020**, *27*, 633–644. [CrossRef]
7. Fox, K.A.A.; Lucas, J.E.; Pieper, K.S.; Bassand, J.P.; Camm, A.J.; Fitzmaurice, D.A.; Goldhaber, S.Z.; Goto, S.; Haas, S.; Hacke, W.; et al. Improved risk stratification of patients with atrial fibrillation: An integrated GARFIELD-AF tool for the prediction of mortality, stroke and bleed in patients with and without anticoagulation. *BMJ Open* **2017**, *7*, e017157. [CrossRef]
8. Horne, B.D.; May, H.T.; Kfoury, A.G.; Renlund, D.G.; Muhlestein, J.B.; Lappe, D.L.; Rasmusson, K.D.; Bunch, T.J.; Carlquist, J.F.; Bair, T.L.; et al. The Intermountain Risk Score (including the red cell distribution width) predicts heart failure and other morbidity endpoints. *Eur. J. Heart Fail.* **2010**, *12*, 1203–1213. [CrossRef]
9. Samaras, A.; Kartas, A.; Akrivos, E.; Fotos, G.; Dividis, G.; Vasdeki, D.; Vrana, E.; Rampidis, G.; Karvounis, H.; Giannakoulas, G.; et al. A novel prognostic tool to predict mortality in patients with atrial fibrillation: The BASIC-AF risk score. *Hellenic. J. Cardiol.* **2021**, *62*, 339–348. [CrossRef]
10. Ahmad, T.; Pencina, M.J.; Schulte, P.J.; O'Brien, E.; Whellan, D.J.; Pina, I.L.; Kitzman, D.W.; Lee, K.L.; O'Connor, C.M.; Felker, G.M. Clinical implications of chronic heart failure phenotypes defined by cluster analysis. *J. Am. Coll. Cardiol.* **2014**, *64*, 1765–1774. [CrossRef]
11. Ahmad, T.; Desai, N.; Wilson, F.; Schulte, P.; Dunning, A.; Jacoby, D.; Allen, L.; Fiuzat, M.; Rogers, J.; Felker, G.M.; et al. Clinical Implications of Cluster Analysis-Based Classification of Acute Decompensated Heart Failure and Correlation with Bedside Hemodynamic Profiles. *PLoS ONE* **2016**, *11*, e0145881. [CrossRef] [PubMed]
12. Burgel, P.R.; Paillasseur, J.L.; Caillaud, D.; Tillie-Leblond, I.; Chanez, P.; Escamilla, R.; Court-Fortune, I.; Perez, T.; Carre, P.; Roche, N.; et al. Clinical COPD phenotypes: A novel approach using principal component and cluster analyses. *Eur. Respir. J.* **2010**, *36*, 531–539. [CrossRef] [PubMed]
13. Antonucci, E.; Poli, D.; Tosetto, A.; Pengo, V.; Tripodi, A.; Magrini, N.; Marongiu, F.; Palareti, G.; Register, S. The Italian START-Register on Anticoagulation with Focus on Atrial Fibrillation. *PLoS ONE* **2015**, *10*, e0124719. [CrossRef] [PubMed]
14. Hennig, C.; Liao, T.F. How to find an appropriate clustering for mixed-type variables with application to socio-economic stratification. *J. R. Stat. Soc.* **2013**, *62*, 309–369. [CrossRef]

15. Link, M.S.; Giugliano, R.P.; Ruff, C.T.; Scirica, B.M.; Huikuri, H.; Oto, A.; Crompton, A.E.; Murphy, S.A.; Lanz, H.; Mercuri, M.F.; et al. Stroke and Mortality Risk in Patients With Various Patterns of Atrial Fibrillation: Results From the ENGAGE AF-TIMI 48 Trial (Effective Anticoagulation With Factor Xa Next Generation in Atrial Fibrillation-Thrombolysis in Myocardial Infarction 48). *Circ. Arrhythm Electrophysiol.* **2017**, *10*, e004267. [CrossRef] [PubMed]
16. Liu, X.; Guo, L.; Xiao, K.; Zhu, W.; Liu, M.; Wan, R.; Hong, K. The obesity paradox for outcomes in atrial fibrillation: Evidence from an exposure-effect analysis of prospective studies. *Obes. Rev.* 2019. [CrossRef]
17. Pastori, D.; Marang, A.; Bisson, A.; Menichelli, D.; Herbert, J.; Lip, G.Y.H.; Fauchier, L. Thromboembolism, mortality, and bleeding in 2,435,541 atrial fibrillation patients with and without cancer: A nationwide cohort study. *Cancer* **2021**, *127*, 2122–2129. [CrossRef]
18. Menichelli, D.; Vicario, T.; Ameri, P.; Toma, M.; Violi, F.; Pignatelli, P.; Pastori, D. Cancer and atrial fibrillation: Epidemiology, mechanisms, and anticoagulation treatment. *Prog. Cardiovasc. Dis.* **2021**, *66*, 28–36. [CrossRef]
19. Pastori, D.; Pignatelli, P.; Sciacqua, A.; Perticone, M.; Violi, F.; Lip, G.Y.H. Relationship of peripheral and coronary artery disease to cardiovascular events in patients with atrial fibrillation. *Int. J. Cardiol.* **2018**, *255*, 69–73. [CrossRef]
20. Aboyans, V.; Ricco, J.B.; Bartelink, M.E.L.; Bjorck, M.; Brodmann, M.; Cohnert, T.; Collet, J.P.; Czerny, M.; De Carlo, M.; Debus, S.; et al. 2017 ESC Guidelines on the Diagnosis and Treatment of Peripheral Arterial Diseases, in collaboration with the European Society for Vascular Surgery (ESVS): Document covering atherosclerotic disease of extracranial carotid and vertebral, mesenteric, renal, upper and lower extremity arteries Endorsed by: The European Stroke Organization (ESO)The Task Force for the Diagnosis and Treatment of Peripheral Arterial Diseases of the European Society of Cardiology (ESC) and of the European Society for Vascular Surgery (ESVS). *Eur. Heart J.* **2017**, *39*, 763–816. [CrossRef]
21. Ntaios, G.; Papavasileiou, V.; Makaritsis, K.; Milionis, H.; Manios, E.; Michel, P.; Lip, G.Y.; Vemmos, K. Statin treatment is associated with improved prognosis in patients with AF-related stroke. *Int. J. Cardiol.* **2014**, *177*, 129–133. [CrossRef] [PubMed]
22. An, Y.; Ogawa, H.; Yamashita, Y.; Ishii, M.; Iguchi, M.; Masunaga, N.; Esato, M.; Tsuji, H.; Wada, H.; Hasegawa, K.; et al. Causes of death in Japanese patients with atrial fibrillation: The Fushimi Atrial Fibrillation Registry. *Eur. Heart J. Qual. Care Clin. Outcomes* **2019**, *5*, 35–42. [CrossRef] [PubMed]
23. Pokorney, S.D.; Piccini, J.P.; Stevens, S.R.; Patel, M.R.; Pieper, K.S.; Halperin, J.L.; Breithardt, G.; Singer, D.E.; Hankey, G.J.; Hacke, W.; et al. Cause of Death and Predictors of All-Cause Mortality in Anticoagulated Patients With Nonvalvular Atrial Fibrillation: Data From ROCKET AF. *J. Am. Heart Assoc.* **2016**, *5*, e002197. [CrossRef] [PubMed]
24. Inohara, T.; Piccini, J.P.; Mahaffey, K.W.; Kimura, T.; Katsumata, Y.; Tanimoto, K.; Inagawa, K.; Ikemura, N.; Ueda, I.; Fukuda, K.; et al. A Cluster Analysis of the Japanese Multicenter Outpatient Registry of Patients With Atrial Fibrillation. *Am. J. Cardiol.* **2019**, *124*, 871–878. [CrossRef] [PubMed]
25. Inohara, T.; Shrader, P.; Pieper, K.; Blanco, R.G.; Thomas, L.; Singer, D.E.; Freeman, J.V.; Allen, L.A.; Fonarow, G.C.; Gersh, B.; et al. Association of of Atrial Fibrillation Clinical Phenotypes With Treatment Patterns and Outcomes: A Multicenter Registry Study. *JAMA Cardiol.* **2018**, *3*, 54–63. [CrossRef] [PubMed]
26. Hindricks, G.; Potpara, T.; Dagres, N.; Arbelo, E.; Bax, J.J.; Blomstrom-Lundqvist, C.; Boriani, G.; Castella, M.; Dan, G.A.; Dilaveris, P.E.; et al. 2020 ESC Guidelines for the diagnosis and management of atrial fibrillation developed in collaboration with the European Association of Cardio-Thoracic Surgery (EACTS). *Eur. Heart J.* **2020**, *42*, 373–498. [CrossRef] [PubMed]
27. Pastori, D.; Menichelli, D.; Violi, F.; Pignatelli, P.; Gregory, Y.H.L.; ATHERO-AF study group. The Atrial fibrillation Better Care (ABC) pathway and cardiac complications in atrial fibrillation: A potential sex-based difference. The ATHERO-AF study. *Eur. J. Intern. Med.* **2021**, *85*, 80–85. [CrossRef]
28. Windgassen, S.; Moss-Morris, R.; Goldsmith, K.; Chalder, T. The importance of cluster analysis for enhancing clinical practice: An example from irritable bowel syndrome. *J. Ment. Health* **2018**, *27*, 94–96. [CrossRef]

Journal of
Personalized
Medicine

Article

Relation of the 'Atrial Fibrillation Better Care (ABC) Pathway' to the Quality of Anticoagulation in Atrial Fibrillation Patients Taking Vitamin K Antagonists

Vanessa Roldán [1], Lorena Martínez-Montesinos [1], Raquel López-Gálvez [2], Lucía García-Tomás [1], Gregory Y. H. Lip [3,4], José Miguel Rivera-Caravaca [2,3,*] and Francisco Marín [2]

[1] Department of Hematology and Clinical Oncology, Hospital General Universitario Morales Meseguer, University of Murcia, Instituto Murciano de Investigación Biosanitaria (IMIB-Arrixaca), 30008 Murcia, Spain; vroldans@um.es (V.R.); l.martinez.montesinos@gmail.com (L.M.-M.); lucia.gt47@gmail.com (L.G.-T.)

[2] Department of Cardiology, Hospital Clínico Universitario Virgen de la Arrixaca, University of Murcia, Instituto Murciano de Investigación Biosanitaria (IMIB-Arrixaca), CIBERCV, 30120 Murcia, Spain; raquellgalvez@gmail.com (R.L.-G.); fcomarino@hotmail.com (F.M.)

[3] Liverpool Centre for Cardiovascular Science, University of Liverpool and Liverpool Heart and Chest Hospital, Liverpool L14 3PE, UK; gregory.lip@liverpool.ac.uk

[4] Department of Clinical Medicine, Aalborg University, 9000 Aalborg, Denmark

* Correspondence: jmrivera429@gmail.com

Abstract: The Atrial Fibrillation Better Care (ABC) pathway was proposed for a more integrated atrial fibrillation (AF) care. We investigated if adherence to the ABC pathway was associated to the quality of anticoagulation control in a cohort of AF outpatients starting vitamin K antagonists (VKAs) between July 2016 and June 2018. Patients were considered adherent to the ABC pathway if they met all of its components. The time in therapeutic range (TTR) was estimated at one year. In total, 1045 patients (51.6% female; median age 77 years; 63% ABC pathway adherent) were included. At one year, 474 (51.6%) of 919 patients with international normalized ratio (INR) data for TTR estimation had a TTR < 65%. Among ABC pathway non-adherent patients, a greater proportion had TRT < 65% (56.4% vs. 43.6%, $p = 0.025$), and TTR < 70% (64.9% vs. 35.1%, $p = 0.033$), with lower mean TTR in non-adherent patients (59.4 $\pm$ 22.3% vs. 63.9 $\pm$ 21.1%; $p = 0.004$). Logistic regression models demonstrated that the ABC pathway adherence in its continuous (aOR: 0.75, 95% CI 0.59–0.96) and categorical (aOR: 0.75, 95% CI 0.57–0.98) forms was independently associated with TTR $\geq$ 65%. In this 'real-world' cohort of AF patients starting VKAs, the ABC pathway adherent patients had better TTR, and more ABC criteria fulfilled increased the probability of achieving good TTR.

Keywords: atrial fibrillation; vitamin K antagonists; time in therapeutic range; Atrial Fibrillation Better Care (ABC) pathway

Citation: Roldán, V.; Martínez-Montesinos, L.; López-Gálvez, R.; García-Tomás, L.; Lip, G.Y.H.; Rivera-Caravaca, J.M.; Marín, F. Relation of the 'Atrial Fibrillation Better Care (ABC) Pathway' to the Quality of Anticoagulation in Atrial Fibrillation Patients Taking Vitamin K Antagonists. *J. Pers. Med.* **2022**, *12*, 487. https://doi.org/10.3390/jpm12030487

Academic Editor: Oscar Campuzano

Received: 4 February 2022
Accepted: 16 March 2022
Published: 17 March 2022

1. Introduction

Atrial fibrillation (AF) is the most common arrhythmia, with a prevalence of ~2% in the overall population and up to 15% in the elderly aged $\geq$80 years old [1]. AF is associated with high morbidity and mortality mainly due to its increased risk of stroke and thromboembolism [2], and oral anticoagulation (OAC, either with vitamin K antagonists [VKAs] or non-vitamin K antagonists [NOACs]) is effective in reducing these risks [3,4]. Despite OAC therapy, cardiovascular complications such as acute coronary syndrome and cardiovascular death are frequent, due to the coexistence of others cardiovascular risk factors such as high blood pressure or diabetes mellitus.

Thus, a more holistic and integrated care approach to managing AF has been proposed, not only focused on stroke prevention, but also efforts to reduce cardiovascular risk factors/comorbidities, including broader approaches such as nurse-led interventions, education and lifestyle modifications (e.g., obesity management, physical exercise and healthy diet) [5,6]. This is proposed in the Atrial Fibrillation Better Care (ABC) pathway, where 'A' refers to 'Avoid stroke'; 'B' refers to 'Better symptom management' and 'C' refers to 'Cardiovascular and comorbidity risk reduction' [7]. Recently, the 2020 European Society of Cardiology guidelines on the management of AF, as well as other international guidelines, have recognized this need for a more integrated care, by including for the first time the ABC pathway as a simplified and concise approach that integrates the care of AF patients across various levels of healthcare professionals and between specialties [8–10].

Despite the increasing use of NOACs, VKAs are still the most commonly used anticoagulants for stroke prevention in AF in several countries. However, the efficacy and tolerability of VKAs depends on the quality of anticoagulant control, as reflected by the mean time in therapeutic range (TTR) of international normalized ratio (INR) 2.0 to 3.0 [11]. Indeed, a high TTR translates into a lower risk of adverse events [12–14]. However, there is no evidence to date whether a more holistic approach to management of AF as reflected by the ABC pathway is associated with better quality of anticoagulation control in VKA users.

In the present study, we investigated the relationship between the ABC pathway and the quality of anticoagulation control in a contemporary cohort of real-world AF patients starting VKA therapy.

2. Materials and Methods

Detailed methods of the present study have been previously published [15]. Briefly, this was a prospective observational cohort study including outpatients newly diagnosed with AF and OAC-naïve attending an anticoagulation clinic of a tertiary hospital (Murcia, Spain), from 1 July 2016 to 30 June 2018. The inclusion criteria were as follows: adult AF patients (i.e., $\geq$18 years old) with documented evidence of AF on ECG and not previously taking OAC for another reason, starting VKAs for the first time. Patients with prosthetic heart valves and severe (mainly rheumatic) valvular AF were excluded. No other exclusion criteria were established.

At baseline, a complete medical history was recorded, including socio-demographic and anthropometric data, comorbidities, concomitant therapies and results of the most recent lab test. Stroke risk (CHA_2DS_2-VASc) and bleeding risk (HAS-BLED) were estimated. We also calculated the SAMe-TT_2R_2 score [Sex, Age (<60 years); Medical history (at least 2 of the following: hypertension, diabetes, coronary artery disease/myocardial infarction, peripheral arterial disease, congestive heart failure, previous stroke, pulmonary disease, hepatic or renal disease); Treatment (interacting drugs, e.g., amiodarone for rhythm control): all 1 point; the current Tobacco use (2 points); and Race (non-Caucasian; 2 points)] as a measure of whether the patient was likely to have good anticoagulation control on VKA [16].

The study protocol was approved by the Ethics Committee from the University Hospital Morales Meseguer (reference: EST: 20/16), and was carried out in accordance with the ethical standards established in the 1964 Declaration of Helsinki and its subsequent amendments. Informed consent was required for participation in this study.

2.1. ABC (Atrial Fibrillation Better Care) Pathway Assessment

The ABC pathway was evaluated according to its original definition, as follows:

'*A' Criterion:* At baseline, a patient would qualify for this criterion if properly prescribed and treated with an OAC. As all patients were included in the context of starting VKA therapy and no previous data about the TTR were available, the 'A' criterion was considered fulfilled if VKA was correctly prescribed according to thromboembolic risk (i.e., CHA_2DS_2-VASc $\geq$ 1 in males or CHA_2DS_2-VASc $\geq$ 2 in females).

'*B' Criterion:* Defined as the presence of symptoms, classified by the European Heart Rhythm Association (EHRA) symptom scale. Any patient with an EHRA score of I (no symptoms) or II (mild symptoms not affecting daily life) qualified for this criterion. Data on symptoms were collected at baseline.

'C' Criterion: Defined as the optimal management/medical treatment of the main cardiovascular comorbidities: hypertension, coronary artery disease, peripheral artery disease, heart failure, stroke/transient ischaemic attack (TIA), and diabetes mellitus. Optimal medical treatment was defined as follows: (i) for hypertension, this was considered controlled if blood pressure <160/90 mmHg was recorded at baseline and treated with appropriate antihypertensive drugs; (ii) for coronary artery disease, treatment with angiotensin-converting enzyme (ACE) inhibitors, beta-blockers, and statins; (iii) for peripheral artery disease or previous stroke/TIA, treatment with statins; (iv) for heart failure, treatment with ACE inhibitors/angiotensin receptor blockers and beta-blockers; and (v) for diabetes mellitus, treatment with insulin or oral antidiabetics. To be included as adherent to 'C' criterion, all main risk factors should have been controlled and/or treated with appropriate drugs.

A patient was considered as fully ABC pathway adherent ('ABC adherent care') if all the three criteria were fulfilled.

2.2. Follow-Up and Study Outcomes

Follow-up was performed according to the standard of care at each routine visit to the outpatient anticoagulation clinic or visits for the anticoagulation control. Medical records and telephone calls were used to obtain the information needed and vital status, if the patient never attends to these visits. No specific interventions and no specific visits were performed for study purposes. During one-year of follow-up, all INR measurements were recorded. The therapeutic range was established at INR 2.0–3.0 according to national and international recommendations.

For the present study, the primary endpoint was the quality of anticoagulation with VKA by using the TTR calculated by the linear interpolation method of Rosendaal at one-year after entry [17]. The linear interpolation proposed by Rosendaal is based on the assumption that moving from one INR to a different one in two determinations separated by a certain number of days occur in a linear way, crossing the difference between the two INR values during those days. Thus, it estimates that the difference between two different INR determinations belonging to different days was produced by increasing or decreasing the INR each day [17]. The secondary endpoint was the quality of anticoagulation with VKA by using the proportion of INRs in range (PINRR, the so-called direct method). This simple method estimates the quality of anticoagulation by taking into account how many INRs are within the therapeutic range (i.e., INR 2.0–3.0) over the total INRs measured [18]. The TTR and PINRR were calculated, excluding the first month of anticoagulation, and we used two cut-off points for the definition of not well controlled VKA therapy (<65% and <70%).

2.3. Statistical Analyses

Continuous variables were expressed as mean $\pm$ standard deviation (SD) or median and interquartile range (IQR) as appropriate, whilst categorical variables were expressed as absolute frequencies and percentages. The Pearson chi-squared test was used to compare proportions, and differences between continuous and categorical variables were assessed using the Mann–Whitney U test or the Student t test, as appropriate. The correlation between the ABC pathway and TTR was tested using the Pearson's r.

Multivariate logistic regression analyses were performed to determine the association between the ABC pathway and the primary/secondary endpoints. A univariate significance level of 0.05 was required to allow a variable into the multivariate model (SLENTRY = 0.05) and a multivariate significance level of 0.05 was required for a variable to stay in the model (SLSTAY = 0.05). Results were reported as adjusted odds ratios (aOR) with a 95% confidence interval (CI).

A p-value < 0.05 was accepted as statistically significant. Statistical analyses were performed using SPSS v. 25.0 (SPSS, Inc., Chicago, IL, USA), and MedCalc v. 16.4.3 (MedCalc Software bvba, Ostend, Belgium) for Windows.

3. Results

Overall, 1254 patients with AF were initially screened and 1064 were included. Of these, 14 patients were lost to follow-up and 5 patients had no all data for the ABC pathway estimation, giving a final study cohort of 1045 patients (Figure 1) (51.6% female; median age 77, IQR 70–83 years) with a median CHA_2DS_2-VASc of 4 (IQR 3–5) and HAS-BLED of 2 (IQR 2–3). Baseline data are summarized in Table 1.

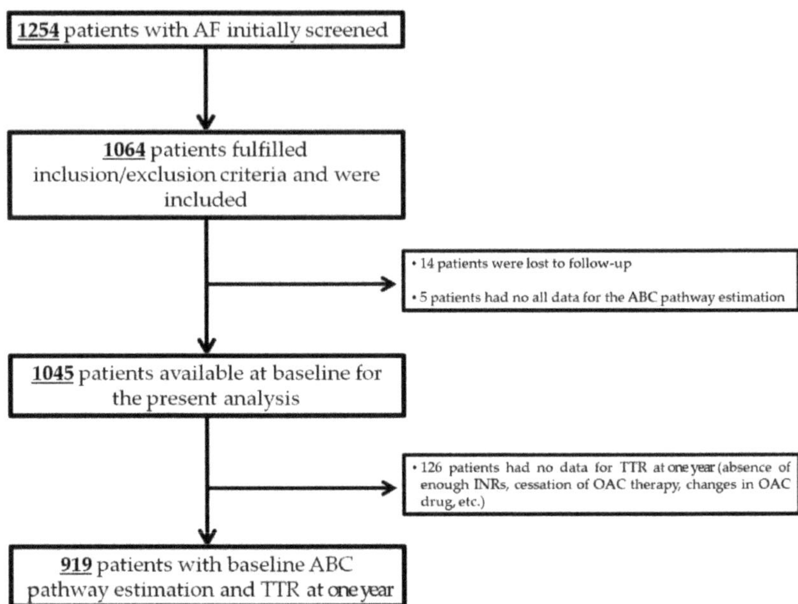

Figure 1. Flow-chart of the study. AF = atrial fibrillation; INR = international normalized ratio; OAC = oral anticoagulation; TTR = time in therapeutic range.

Table 1. Baseline clinical characteristics.

	N = 1045
Demographic	
Male sex, n (%)	506 (48.4)
Age (years), median (IQR)	77 (70–83)
BMI (kg/m^2), median (IQR)	30.0 (26.8–33.3)
Comorbidities, n (%)	
Hypertension	874 (83.6)
Diabetes mellitus	393 (37.6)
Heart failure	261 (25.0)
History of stroke/TIA/thromboembolism	162 (15.5)
Renal impairment	197 (18.9)
Coronary artery disease	190 (18.2)
Peripheral artery disease	66 (6.3)
Hypercholesterolemia	608 (58.2)
Current smoking habit	157 (15.0)
Current alcohol consumption	71 (6.8)
History of previous bleeding	173 (16.6)
COPD/OSAH	230 (22.0)
Hepatic disease	68 (6.5)
Concomitant malignant disease	150 (14.4)

Table 1. *Cont.*

	N = 1045
Concomitant treatment, n (%)	
Antiarrhythmics	214 (20.5)
Calcium antagonist	317 (30.3)
Beta-blockers	723 (69.2)
Statins	555 (53.1)
Diuretics	571 (54.6)
Antiplatelet therapy	256 (24.5)
ACE inhibitors	255 (24.4)
Angiotensin II receptor blockers	456 (43.6)
Oral antidiabetics/insulin	279 (26.7)
CHA_2DS_2-VASc, median (IQR)	4 (3–5)
HAS-BLED, median (IQR)	2 (2–3)
SAMe-TT_2R_2, median (IQR)	1 (1–2)

ACE inhibitors = angiotensin-converting enzyme inhibitors; COPD/OSAH = chronic obstructive pulmonary disease/obstructive sleep apnoea/hypopnoea; BMI = body mass index; IQR = interquartile range; TIA = transient ischemic attack.

With regards to the ABC pathway, 1017 (97.3%) of patients fulfilled the "A" criterion at baseline; 890 (85.2%) fulfilled the "B" criterion; and 809 (77.4%) fulfilled the "C" criterion. Overall, 32 (3.1%) were adherent to one criterion, 355 (34.0%) were adherent to two criteria, and 658 (63.0%) were adherent to all three criteria. Thus, 658 (63%) were categorized as adherent to the ABC pathway at baseline, whereas 387 (37%) were considered not adherent.

However, enough INR data for TTR estimation was available in 919 patients at one-year of follow-up (Figure 1). The mean TTR of these patients was 62.3% ± 21.7%; 474 (51.6%) of them did not achieve a TTR over 65% (mean TTR 45.9% ± 16.2%) and 555 (60.4%) did not achieve a TTR over 70% (mean TTR 49.0% ± 16.8%). Among those non-adherent to the ABC pathway at baseline, a higher proportion presented a TTR < 65% than TTR $\geq$ 65% (56.4% vs. 43.6%, $p = 0.023$), as well as a higher proportion had TTR < 70% compared to TTR $\geq$ 70% (64.9% vs. 35.1%, $p = 0.031$) at one year. Of note, the mean TTR was lower in those non-adherent to the ABC pathway compared to ABC-adherent patients (59.4% ± 22.3% vs. 63.9% ± 21.1%; $p = 0.002$).

3.1. ABC Pathway Adherence and Quality of Anticoagulation Control

The ABC pathway as a continuous parameter and the TTR were significantly correlated ($p < 0.001$) (Figure 2). A logistic regression model showed that the ABC pathway in its continuous form was associated with the quality of anticoagulation control, even after adjusting for the SAMe-TT_2R_2 score. A greater number of ABC pathway criteria fulfilled was independently associated with a TTR $\geq$ 65% (adjusted OR 0.75, 95% CI 0.59–0.96, $p = 0.020$), and with a TTR $\geq$ 70% (aOR 0.73, 95% CI 0.57–0.94, $p = 0.015$).

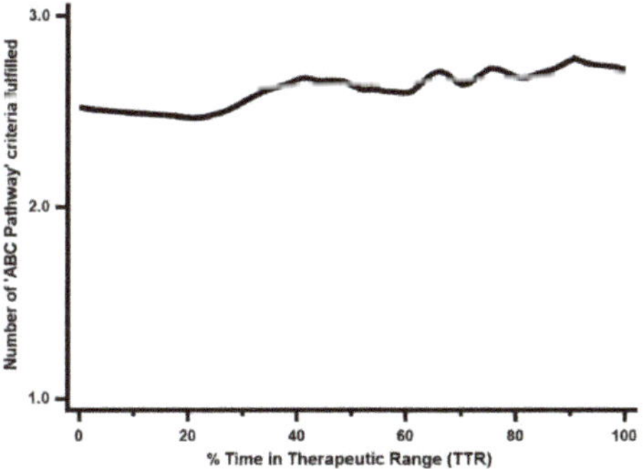

Figure 2. Scatter diagram showing the graphical correlation of the ABC pathway and TTR.

Patients adherent to the three ABC pathway criteria had a significantly higher probability of achieving TTR $\geq$ 65% (aOR 0.41, 95% CI 0.18–0.95, p = 0.038) when compared to patients adherent to one or two criteria only. Fulfilling the "C" criterion alone was also significantly related to TTR $\geq$ 65% (aOR 0.67, 95% CI 0.49–0.91, p = 0.010). Importantly, the results were consistent when using the TTR $\geq$ 70% as the cut-off (Table 2).

Table 2. Association of ABC components with quality of anticoagulation therapy by different thresholds.

	TTR < 65%		TTR < 70%	
	aOR (95% CI)	***p*-Value**	**aOR (95% CI)**	***p*-Value**
ABC pathway (1 criterion)	Ref.		Ref.	
ABC pathway (2 criteria)	0.51 (0.22–1.20)	0.125	0.30 (0.10–0.88)	0.028
ABC pathway (3 criteria)	0.41 (0.18–0.95)	0.038	0.25 (0.08–0.72)	0.010
ABC pathway (A criterion)	0.56 (0.24–1.33)	0.191	0.42 (0.16–1.15)	0.092
ABC pathway (B criterion)	1.02 (0.70–1.48)	0.923	0.99 (0.68–1.46)	0.974
ABC pathway (C criterion)	0.67 (0.49–0.91)	0.010	0.66 (0.48–0.92)	0.013

aOR = adjusted odds ratio; CI = confidence interval; TTR = time in therapeutic range.

In its categorical form (i.e., adherent vs. non-adherent), the ABC pathway was significantly associated with the quality of anticoagulation control in the model adjusted for SAMe-TT$_2$R$_2$. Patients categorized as adherent to the ABC pathway had a higher probability of TTR $\geq$ 65% (aOR 0.75, 95% CI 0.57–0.98, p = 0.039), although a non-significant trend was seen when the cut-off point for TTR was 70% (aOR 0.77, 95% CI 0.58–1.02, p = 0.065) (Figure 3).

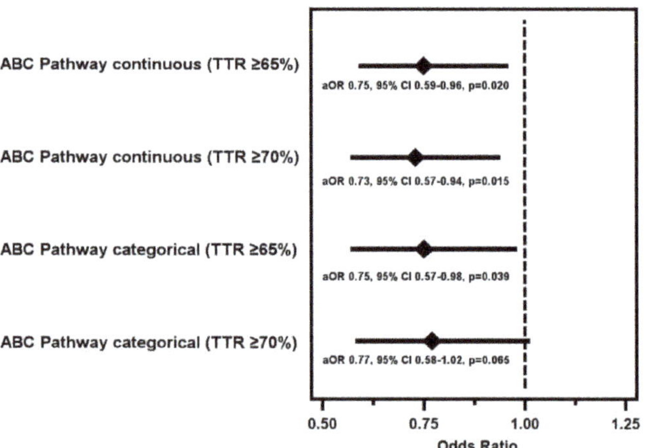

Figure 3. Probability of poor TTR in relation to the ABC pathway.

3.2. ABC Pathway Compliance and PINRR

With regard to the secondary outcome, after adjusting for the SAMe-TT$_2$R$_2$ score, the ABC pathway was independently associated with a PINRR $\geq$ 65%, with an aOR of 0.76 (95% CI 0.59–0.98, p = 0.044) for the continuous form and an aOR of 0.72 (95% CI 0.54–0.97, p = 0.029) for the categorical form (adherent vs. non-adherent). This was not statistically significant for PINRR $\geq$ 70% (aOR 0.76, 95% CI 0.56–1.02, p = 0.067 and aOR 0.75, 95% CI 0.54–1.04, p = 0.087; respectively).

4. Discussion

In this study, we demonstrated that an integrated management according to the 'ABC pathway' could lead to better anticoagulation therapy with VKAs in terms of TTR. This is central since the efficacy and safety of VKA depends on the quality of anticoagulant control, as reflected by the average TTR of INRs 2.0 to 3.0. Various studies have shown how a high TTR translates into a lower risk of stroke and bleeding while on OAC [12,13,19]. The maintenance dose of VKA is influenced by many different factors, including race, dietary vitamin K intake, comorbidities (e.g., liver disease and acute illness), or whether the patient may be taking interacting drugs [20]. The average individual TTR range generally increases over time, but even in very well and experienced patients, the TTR can decrease during follow-up [3]. In this context, drug adherence is important in TTR maintenance; indeed, non-adherence and low anticoagulation control levels are associated [21,22]. Adherence is also linked to a better knowledge of the patient about its disease and the treatment. In patients taking warfarin in an anticoagulation clinic who completed a questionnaire survey about knowledge, satisfaction and concerns regarding warfarin treatment, those with better knowledge and higher satisfaction were those with higher warfarin adherence and better INR control [23–25].

The SAMe-TT$_2$R$_2$ score was proposed to identify those AF patients that would be less likely to do well with good anticoagulation control on VKAs [16]. As the SAMe-TT$_2$R$_2$ score includes several comorbidities/cardiovascular risk factors, it not only predicts poor anticoagulation control but also adverse events [26]. Indeed, the SAMe-TT$_2$R$_2$ score sums up the influence of comorbidities and adjunctive treatment on the response to VKA treatment. The ABC pathway goes beyond this, since it demonstrates that correctly managed AF patients and the ABC-adherent might have higher TTR, beyond the presence of comorbidities alone. In this setting, we demonstrated in the present study that the ABC pathway is associated with well-managed VKA treatment, independently of the SAMe-TT$_2$R$_2$ score.

Despite the limitations of VKAs (narrow therapeutic range, multiple food and drugs interactions), investing in education and counselling would improve the quality of anticoagulation control amongst VKA-anticoagulated patients [27]. One pilot study showed that even a brief educational intervention can help to improve the knowledge about anticoagulation therapy for AF, focused on patient's knowledge of the target INR range and factors that may affect INR levels [28].

Since its first publication, several studies have demonstrated that ABC pathway-adherent-patients have better outcomes that those who are not adherent [29–31]. In a systematic review including eight studies and more than 285,000 AF patients, there was a pooled prevalence of 21% who were ABC-adherent patients [32]. Importantly, adherence to the ABC pathway was associated with a reduction in the risk of major adverse outcomes [32]. However, most of the evidence about the ABC pathway today derives from NOAC-treated AF cohorts. Nevertheless, a low TTR is associated with worse outcomes [33], including higher risk of mortality [13]. Thus, TTR and clinical outcomes are intimately related. In this context, our study addresses the fact that patients who fulfilled the ABC pathway showed a higher TTR value perhaps because of better clinical management. AF patients are considered as a clinical complex group with multiple comorbidities, polypharmacy and multiple cardiovascular events, so a holistic approach would result in better management and a reduction not only for worse clinical outcomes but also for higher quality of anticoagulation control with VKA therapy. Despite NOACs have changed the landscape of OAC in AF patients, there are still regional differences in the prescription of these drugs (for example, in Spain, whereby NOACs are only reimbursed if AF patients fulfill very specific conditions). There are also patients for whom NOACs are contraindicated, for example, AF patients with a mechanical heart valve, who need good anticoagulation control with VKAs. The PLECTRUM study, which includes 2111 patients with mechanical heart valves, showed that hypertension, diabetes and heart failure were independently associated with a low TTR, providing clinical evidence that comorbidities directly affect anticoagulant management and better INR control [34].

In summary, adherence with ABC pathway management is not only associated with a lower number of cardiovascular events but also results in better anticoagulation control, which in turn is associated with a better prognosis. This strategy is useful for both types of

OACs, as NOACs require a close follow-up to ensure good adherence, while VKAs need a high TTR for optimal clinical benefit.

Limitations

Our study has some limitations. The main limitation of the study lies in its observational nature, with a Caucasian-based population and single centre design, performed in a unique anticoagulation clinic. It has been previously demonstrated that anticoagulation clinics have better results in terms of the TTR achieved [35], so data from other anticoagulation management models should be explored. We included patients who were VKA-naïve, which has two drawbacks: first, the start of anticoagulation with VKA is a complex period, with a greater number of adverse events; and second, not all patients reach the therapeutic range at the same time.

5. Conclusions

In this 'real world' prospective cohort of AF patients starting VKA therapy, management adherent to the ABC pathway was related to the quality of anticoagulation control at one year. Adherent patients to the ABC pathway had better TTR, and more ABC criteria fulfilled were associated with higher probability of achieving good TTR.

Author Contributions: Conceptualization, V.R., J.M.R.-C. and G.Y.H.L.; methodology, V.R. and J.M.R.-C. formal analysis, J.M.R.-C.; investigation, V.R., L.M.-M., L.G.-T., R.L.-G. and J.M.R.-C.; data acquisition, V.R., L.M.-M. and J.M.R.-C.; data curation, J.M.R.-C.; writing—original draft preparation, V.R.; writing—review and editing, F.M., G.Y.H.L. and J.M.R.-C. funding acquisition, V.R., and F.M. All authors have read and agreed to the published version of the manuscript.

Funding: This work was supported by the Spanish Ministry of Economy, Industry, and Competitiveness, through the Instituto de Salud Carlos III after independent peer review (research grant: PI17/01375 co-financed by the European Regional Development Fund) and group CB16/11/00385 from CIBERCV.

Institutional Review Board Statement: The study was conducted in accordance with the Declaration of Helsinki, and approved by the Ethics Committee of University Hospital Morales Meseguer (reference: EST: 20/16, date of approval: 22 June 2016).

Informed Consent Statement: Informed consent was obtained from all subjects involved in the study.

Data Availability Statement: Data cannot be shared for ethical/privacy reasons.

Conflicts of Interest: G.Y.H.L.: Consultant and speaker for BMS/Pfizer, Boehringer Ingelheim, and Daiichi-Sankyo. No fees are received personally. There is nothing to disclose for other authors.

References

1. Kornej, J.; Börschel, C.S.; Benjamin, E.J.; Schnabel, R.B. Epidemiology of Atrial Fibrillation in the 21st Century: Novel Methods and New Insights. *Circ. Res.* **2020**, *127*, 4–20. [CrossRef]
2. Lip, G.Y.; Freedman, B.; De Caterina, R.; Potpara, T.S. Stroke prevention in atrial fibrillation: Past, present and future. Comparing the guidelines and practical decision-making. *Thromb. Haemost.* **2017**, *117*, 1230–1239. [CrossRef] [PubMed]
3. Rivera-Caravaca, J.M.; Roldán, V.; Esteve-Pastor, M.A.; Valdés, M.; Vicente, V.; Lip, G.Y.H.; Marín, F. Cessation of oral anticoagulation is an important risk factor for stroke and mortality in atrial fibrillation patients. *Thromb. Haemost.* **2017**, *117*, 1448–1454. [CrossRef] [PubMed]
4. Chan, N.; Sobieraj-Teague, M.; Eikelboom, J.W. Direct oral anticoagulants: Evidence and unresolved issues. *Lancet* **2020**, *396*, 1767–1776. [CrossRef]
5. Stevens, D.; Harrison, S.L.; Kolamunnage-Dona, R.; Lip, G.Y.H.; Lane, D.A. The Atrial Fibrillation Better Care pathway for managing atrial fibrillation: A review. *Europace* **2021**, *23*, 1511–1527. [CrossRef]
6. Gallagher, C.; Hendriks, J.M.; Nyfort-Hansen, K.; Sanders, P.; Lau, D.H. Integrated care for atrial fibrillation: The heart of the matter. *Eur. J. Prev. Cardiol.* **2020**. [CrossRef]
7. Lip, G.Y.H. The ABC pathway: An integrated approach to improve AF management. *Nat. Rev. Cardiol.* **2017**, *14*, 627–628. [CrossRef]
8. Hindricks, G.; Potpara, T.; Dagres, N.; Arbelo, E.; Bax, J.J.; Blomström-Lundqvist, C.; Boriani, G.; Castella, M.; Dan, G.-A.; Dilaveris, P.E.; et al. 2020 ESC Guidelines for the diagnosis and management of atrial fibrillation developed in collaboration with the European Association of Cardio-Thoracic Surgery (EACTS). *Eur. Heart J.* **2021**, *42*, 373–498. [CrossRef]

9. Lip, G.Y.; Banerjee, A.; Boriani, G.; Chiang, C.E.; Fargo, R.; Freedman, B.; Lane, D.A.; Ruff, C.T.; Turakhia, M.; Werring, D.; et al. Antithrombotic Therapy for Atrial Fibrillation: CHEST Guideline and Expert Panel Report. *Chest* **2018**, *154*, 1121–1201. [CrossRef]

10. Chao, T.-F.; Joung, B.; Takahashi, Y.; Lim, T.W.; Choi, E.-K.; Chan, Y.-H.; Guo, Y.; Sriratanasathavorn, C.; Oh, S.; Okumura, K.; et al. 2021 Focused Update Consensus Guidelines of the Asia Pacific Heart Rhythm Society on Stroke Prevention in Atrial Fibrillation: Executive Summary. *Thromb. Haemost.* **2021**, *122*, 20–47. [CrossRef]

11. Van den Ham, H.A.; Klungel, O.H.; Leufkens, H.G.M.; Van Staa, T.P.; Van Staa, T. The patterns of anticoagulation control and the risk of stroke, bleeding and mortality in patients with non-valvular atrial fibrillation. *J. Thromb. Haemost.* **2013**, *11*, 107–115. [CrossRef] [PubMed]

12. Pastori, D.; Pignatelli, P.; Cribari, F.; Carnevale, R.; Saliola, M.; Violi, F.; Lip, G.Y. Time to therapeutic range (TtTR), anticoagulation control, and cardiovascular events in vitamin K antagonists–naive patients with atrial fibrillation. *Am. Heart J.* **2018**, *200*, 32–36. [CrossRef] [PubMed]

13. Rivera-Caravaca, J.; Roldán, V.; Esteve-Pastor, M.A.; Valdés, M.; Vicente, V.; Marin, F.; Lip, G.Y. Reduced Time in Therapeutic Range and Higher Mortality in Atrial Fibrillation Patients Taking Acenocoumarol. *Clin. Ther.* **2018**, *40*, 114–122. [CrossRef] [PubMed]

14. Prochaska, J.H.; Hausner, C.; Nagler, M.; Göbel, S.; Eggebrecht, L.; Panova-Noeva, M.; Arnold, N.; Lauterbach, M.; Bickel, C.; Michal, M.; et al. Subtherapeutic Anticoagulation Control under Treatment with Vitamin K-Antagonists—Data from a Specialized Coagulation Service. *Thromb. Haemost.* **2019**, *119*, 1347–1357. [CrossRef]

15. Rivera-Caravaca, J.M.; Marín, F.; Esteve-Pastor, M.A.; Gálvez, J.; Lip, G.Y.; Vicente, V.; Roldán, V. Murcia atrial fibrillation project II: Protocol for a prospective observational study in patients with atrial fibrillation. *BMJ Open.* **2019**, *9*, e033712. [CrossRef]

16. Apostolakis, S.; Sullivan, R.M.; Olshansky, B.; Lip, G.Y. Factors Affecting Quality of Anticoagulation Control Among Patients With Atrial Fibrillation on Warfarin: The SAMe-TT$_2$R$_2$ score. *Chest* **2013**, *144*, 1555–1563. [CrossRef]

17. Rosendaal, F.R.; Cannegieter, S.C.; Van Der Meer, F.J.; Briët, E. A method to determine the optimal intensity of oral anticoagulant therapy. *Thromb. Haemost.* **1993**, *69*, 236–239. [CrossRef]

18. Chan, P.-H.; Li, W.-H.; Hai, J.-J.; Chan, E.W.; Wong, I.C.; Tse, H.-F.; Lip, G.Y.; Siu, C.-W. Time in Therapeutic Range and Percentage of International Normalized Ratio in the Therapeutic Range as a Measure of Quality of Anticoagulation Control in Patients with Atrial Fibrillation. *Can. J. Cardiol.* **2016**, *32*, 1247.e23–1247.e28. [CrossRef]

19. Gallagher, A.M.; Setakis, E.; Plumb, J.M.; Clemens, A.; van Staa, T.-P. Risks of stroke and mortality associated with suboptimal anticoagulation in atrial fibrillation patients. *Thromb. Haemost.* **2011**, *106*, 968–977. [CrossRef]

20. Garcia, D.A.; Lopes, R.D.; Hylek, E.M. New-onset atrial fibrillation and warfarin initiation: High risk periods and implications for new antithrombotic drugs. *Thromb. Haemost.* **2010**, *104*, 1099–1105.

21. Marcatto, L.R.; Sacilotto, L.; Tavares, L.C.; Facin, M.; Olivetti, N.; Strunz, C.M.C.; Darrieux, F.; Scanavacca, M.; Krieger, J.E.; Pereira, A.C.; et al. Pharmaceutical Care Increases Time in Therapeutic Range of Patients with Poor Quality of Anticoagulation with Warfarin. *Front. Pharmacol.* **2018**, *9*, 1052. [CrossRef] [PubMed]

22. Proietti, M.; Lane, D.A. The Compelling Issue of Nonvitamin K Antagonist Oral Anticoagulant Adherence in Atrial Fibrillation Patients: A Systematic Need for New Strategies. *Thromb. Haemost.* **2020**, *120*, 369–371. [CrossRef] [PubMed]

23. Rivera-Caravaca, J.M.; Esteve-Pastor, M.A.; Roldán, V.; Marín, F.; Lip, G.Y. Non-vitamin K antagonist oral anticoagulants: Impact of non-adherence and discontinuation. *Expert Opin. Drug Saf.* **2017**, *16*, 1051–1062. [CrossRef] [PubMed]

24. Wang, Y.; Kong, M.C.; Lee, L.H.; Ng, H.J.; Ko, Y. Knowledge, satisfaction, and concerns regarding warfarin therapy and their association with warfarin adherence and anticoagulation control. *Thromb. Res.* **2014**, *133*, 550–554. [CrossRef]

25. Kim, J.H.; Kim, G.S.; Kim, E.J.; Park, S.; Chung, N.; Chu, S.H. Factors Affecting Medication Adherence and Anticoagulation Control in Korean Patients Taking Warfarin. *J. Cardiovasc. Nurs.* **2011**, *26*, 466–474. [CrossRef]

26. Gallego, P.; Roldán, V.; Marin, F.; Gálvez, J.; Valdés, M.; Vicente, V.; Lip, G.Y. SAMe-TT2R2 Score, Time in Therapeutic Range, and Outcomes in Anticoagulated Patients with Atrial Fibrillation. *Am. J. Med.* **2014**, *127*, 1083–1088. [CrossRef]

27. Clarkesmith, D.; Pattison, H.; Lip, G.Y.H.; Lane, D.A. Educational Intervention Improves Anticoagulation Control in Atrial Fibrillation Patients: The TREAT Randomised Trial. *PLoS ONE* **2013**, *8*, e74037. [CrossRef]

28. Lane, D.A.; Ponsford, J.; Shelley, A.; Sirpal, A.; Lip, G.Y. Patient knowledge and perceptions of atrial fibrillation and anticoagulant therapy: Effects of an educational intervention programme: The West Birmingham Atrial Fibrillation Project. *Int. J. Cardiol.* **2006**, *110*, 354–358. [CrossRef]

29. Pastori, D.; Pignatelli, P.; Menichelli, D.; Violi, F.; Lip, G.Y. Integrated Care Management of Patients with Atrial Fibrillation and Risk of Cardiovascular Events: The ABC (Atrial fibrillation Better Care) Pathway in the ATHERO-AF Study Cohort. *Mayo Clin. Proc.* **2019**, *94*, 1261–1267. [CrossRef]

30. Proietti, M.; Lip, G.Y.H.; Laroche, C.; Fauchier, L.; Marin, F.; Nabauer, M.; Potpara, T.; Dan, G.-A.; Kalarus, Z.; Tavazzi, L.; et al. Relation of outcomes to ABC (Atrial Fibrillation Better Care) pathway adherent care in European patients with atrial fibrillation: An analysis from the ESC-EHRA EORP Atrial Fibrillation General Long-Term (AFGen LT) Registry. *Europace* **2021**, *23*, 174–183. [CrossRef]

31. Yoon, M.; Yang, P.-S.; Jang, E.; Yu, H.T.; Kim, T.-H.; Uhm, J.-S.; Kim, J.-Y.; Sung, J.-H.; Pak, H.-N.; Lee, M.-H.; et al. Improved Population-Based Clinical Outcomes of Patients with Atrial Fibrillation by Compliance with the Simple ABC (Atrial Fibrillation Better Care) Pathway for Integrated Care Management: A Nationwide Cohort Study. *Thromb. Haemost.* **2019**, *119*, 1695–1703. [CrossRef] [PubMed]

32. Romiti, G.F.; Pastori, D.; Rivera-Caravaca, J.M.; Ding, W.Y.; Gue, Y.X.; Menichelli, D.; Gumprecht, J.; Kozieł, M.; Yang, P.-S.; Guo, Y.; et al. Adherence to the 'Atrial Fibrillation Better Care' Pathway in Patients with Atrial Fibrillation: Impact on Clinical Outcomes—A Systematic Review and Meta-Analysis of 285,000 Patients. *Thromb. Haemost.* **2021**, *122*, 406–414. [CrossRef] [PubMed]
33. Wan, Y.; Heneghan, C.; Perera, R.; Roberts, N.; Hollowell, J.; Glasziou, P.; Bankhead, C.; Xu, Y. Anticoagulation Control and Prediction of Adverse Events in Patients with Atrial Fibrillation: A systematic review. *Circ. Cardiovasc. Qual. Outcomes* **2008**, *1*, 84–91. [CrossRef] [PubMed]
34. Pastori, D.; Lip, G.Y.H.; Poli, D.; Antonucci, E.; Rubino, L.; Menichelli, D.; Saliola, M.; Violi, F.; Palareti, G.; Pignatelli, P.; et al. Determinants of low-quality warfarin anticoagulation in patients with mechanical prosthetic heart valves. The nationwide PLECTRUM study. *Br. J. Haematol.* **2020**, *190*, 588–593. [CrossRef]
35. Van Walraven, C.; Jennings, A.; Oake, N.; Fergusson, D.; Forster, A.J. Effect of Study Setting on Anticoagulation Control: A systematic review and metaregression. *Chest* **2006**, *129*, 1155–1166. [CrossRef]

Journal of
*Personalized
Medicine*

Article

The Dissimilar Impact in Atrial Substrate Modification of Left and Right Pulmonary Veins Isolation after Catheter Ablation of Paroxysmal Atrial Fibrillation

Aikaterini Vraka [1], Vicente Bertomeu-González [2], Lorenzo Fácila [3], José Moreno-Arribas [2], Raúl Alcaraz [4] and José J. Rieta [1,*]

[1] BioMIT.org, Electronic Engineering Department, Universitat Politecnica de Valencia, 46022 Valencia, Spain; aivra@upv.es

[2] Cardiology Department, Saint John's University Hospital, 03550 Alicante, Spain; vbertog@gmail.com (V.B.-G.); jomoreno@gmail.com (J.M.-A.)

[3] Cardiology Department, General University Hospital Consortium of Valencia, 46014 Valencia, Spain; lfacila@gmail.com

[4] Research Group in Electronic, Biomedical and Telecommunication Engineering, University of Castilla-La Mancha, 16071 Cuenca, Spain; raul.alcaraz@uclm.es

* Correspondence: jjrieta@upv.es

Abstract: Since the discovery of pulmonary veins (PVs) as foci of atrial fibrillation (AF), the commonest cardiac arrhythmia, investigation revolves around PVs catheter ablation (CA) results. Notwithstanding, CA process itself is rather neglected. We aim to decompose crucial CA steps: coronary sinus (CS) catheterization and the impact of left and right PVs isolation (LPVI, RPVI), separately. We recruited 40 paroxysmal AF patients undergoing first-time CA and obtained five-minute lead II and bipolar CS recordings during sinus rhythm (SR) before CA (**B**), after LPVI (**L**) and after RPVI (**R**). Among others, duration, amplitude and atrial-rate variability (ARV) were calculated for P-waves and CS local activation waves (LAWs). LAWs features were compared among CS channels for reliability analysis. P-waves and LAWs features were compared after each ablation step (**B, L, R**). CS channels: amplitude and area were different between distal/medial ($p \leq 0.0014$) and distal/mid-proximal channels ($p \leq 0.0025$). Medial and distal showed the most and least coherent values, respectively. Correlation was higher in proximal ($\geq$93%) than distal ($\leq$91%) areas. P-waves: duration was significantly shortened after LPVI (after **L**: $p = 0.0012$, -13.30%). LAWs: insignificant variations. ARV modification was more prominent in LAWs (**L**: $>+73.12$%, $p \leq 0.0480$, **R**: <-33.94%, $p \leq 0.0642$). Medial/mid-proximal channels are recommended during SR. CS LAWs are not significantly affected by CA but they describe more precisely CA-induced ARV modifications. LPVI provokes the highest impact in paroxysmal AF CA, significantly modifying P-wave duration.

Keywords: atrial fibrillation; catheter ablation; coronary sinus; catheter channels; P-waves; local activation waves; left pulmonary veins; heart rate variability

Citation: Vraka, A.; Bertomeu-González, V.; Fácila, L.; Moreno-Arribas, J.; Alcaraz, R.; Rieta, J.J. The Dissimilar Impact in Atrial Substrate Modification of Left and Right Pulmonary Veins Isolation after Catheter Ablation of Paroxysmal Atrial Fibrillation. *J. Pers. Med.* **2022**, *12*, 462. https://doi.org/10.3390/jpm12030462

Academic Editor: José Miguel Rivera-Caravaca

Received: 4 February 2022
Accepted: 7 March 2022
Published: 14 March 2022

1. Introduction

Atrial fibrillation (AF) is the prevailing cardiac arrhythmia in the western world. Prolonged lifespan and the connection with a plenty of other comorbidities contribute to the ever-growing AF incidence. Health and economic burden caused by AF alert the need for thorough investigation on its pathophysiology [1]. AF springs principally from pulmonary veins (PVs) [2] and propagates through cardiac structures [3]. The main mechanism assisting the AF propagation is structural remodeling and fibrosis is especially contributing to the alteration of the cardiac anatomy, causing conduction heterogeneity, hence favoring the AF perpetuation [1,4]. Although conduction heterogeneity is more prominent during AF, the anatomical substrate can still be present for both atria even when patients are in sinus rhythm (SR) [5–7]. As PVs are the main AF foci, their electrical isolation, called catheter ablation (CA), is the star AF treatment [1,8]. Despite the high CA success rates for paroxysmal AF patients, persistent AF cases often require the CA of additional cardiac structures that trigger or propagate the AF activity, known as non-PV triggers [1,3,4,9–12].

Many techniques exist to localize non-PV triggers, with complex fractionated atrial electrograms (CFAEs) during AF [4,13,14] or low voltage electrograms (EGMs) during SR [4,15,16] being two of the most established ones. A combination of both techniques along with highly proportioned EGM fractionation has recently indicated sites showing fibrosis, with a high correlation between these areas in AF and SR [6]. Nevertheless, the effect of CA on additional non-PV triggers remains quite controversial. Evidence shows that additional ablation of these sites offers little or no improved results with respect to single PVs ablation [17–19]. It remains unclear, however, whether failure of additional CA applications to provide significant improvement in termination of AF stems from the incapacity of CA on sites other than PVs to terminate AF or from a vague and unclear definition of areas in need of ablation due to highly complex EGMs, thus highlighting the need for more reliable algorithms able to properly evaluate the atrial substrate [6,14].

So far, CA outcome on paroxysmal AF patients is primarily assessed from the analysis of the characteristics of P-waves, which represent the activation of the atria or heart-rate (HR) variability (HRV) analysis, which assesses the ventricular response, controlled by the autonomous nervous system (ANS). P-wave duration (PWD) is the most popular P-wave feature, reflecting the overall time that the wavefront needs to be propagated throughout the atria. Existence of prolonged or short PWD is considered an indicator of AF recurrence in paroxysmal or persistent AF patients, caused by conduction heterogeneity and scarring or shortening of the atrial refractory period, respectively [20–25]. PWD shortening is connected with the elimination of the conduction heterogeneity, hence being a favorable CA marker, while it is the second P-wave part, corresponding to left atrial depolarization, that is mainly modified after CA, possibly due to vicinity with PVs, the main object of CA [26–29]. PWD analysis goes beyond CA procedures, with application in studies predicting the AF occurrence or the risk for higher AF burden after pacemaker implantation [25,30,31].

Apart from PWD, P-wave dispersion, amplitude, area or P-wave to R-peak interval are popular features utilized to predict AF recurrence [26,27,32,33]. P-wave analysis has been additionally applied to frequency domain in order to discern among healthy and AF subjects [34]. HRV is a marker of fine tuning of ANS, which consists of sympathetic and parasympathetic systems and controls sinus rhythm. Evidence shows that people with low HRV are susceptible to AF [35–37]. Energy delivered by radiofrequency (RF) CA (RFCA) can disturb the balance between sympathetic and parasympathetic systems, by stimulating the former and leading to temporary withdrawal of the latter, hence causing HRV attenuation, which in turn has been associated with AF recurrence [38–40].

The number of studies and techniques aiming to analyze the CA effect on the atrial substrate is endless. At the same time, critical CA steps and their impact on the atrial substrate alteration is a rather neglected analysis field. Firstly, the aforementioned studies observing P-wave and HRV alterations only employ recordings acquired before and after CA. This postulates the theory of a uniform impact of left (LPVI) and right PVs isolation (RPVI). It should be considered, though, the possibility of each PV side playing a different role in atrial substrate alteration and hence, in AF activity, a conjecture that can be easily verified by employing signals recorded in between the ablation of LPVI and RPVI, which are already available in the recordings of any electrophysiology laboratory during stepwise CA procedures.

Coronary sinus' (CS) strategical position between left (LA) and right atrium (RA) allows the detection of non-PV triggers and PV reconnection gaps throughout the atria during CA procedures via CS catheterization [41–49]. Despite its extensive use as a CA reference, whether CS analysis could provide reliable information with respect to the AF substrate modification or which channels of CS catheter are the most appropriate for the analysis are two vital issues that remain unexplored. During CS cannulation, the most proximal pair of electrodes (9–10) is placed close to RA and the most distal pair (1–2) close to LA [44,47]. Notwithstanding, CS catheterization may be rather challenging due to variable CS anatomy and shape, aggrravated by myocardial contraction or the existence of CS dilation, factors that can lead to unstable recordings, especially from the distal tip of the catheter [50–53]. Additionally, anatomical alterations of the, adjacent to CS extremes, mitral annulus across the cardiac cycle in SR may affect furthermore the stability of recordings acquired from distal and proximal electrodes of the CS catheter [54,55]. Considering the aforementioned factors, information recorded across the CS catheter electrodes could vary significantly and the choice of the appropriate channel recruited for the analysis should be made with extreme caution.

The present work aims to elucidate the aforeposed issues regarding the CA procedure, in order to arise the understanding on the mechanisms of important CA steps and their interaction with the CA result. In the first place, the ability of CS channels to describe with the highest precision possible the AF dynamics during SR is assessed and the most and least recommended CS channels are defined. Afterwards, the relevance of CS in substrate modification evaluation due to CA is investigated via analysis of features traditionally applied to ECG recordings and cross-referenced by P-waves and HRV analysis of the ECGs. Finally, the evolution of P-waves and CS LAWs after isolation of either sides of PVs is tracked in order to define the PV side that has the highest impact in atrial substrate modification due to CA.

The manuscript is organized as follows. Section 2 briefly describes the database recruited for the analysis, as well as the preprocessing and analysis steps. Section 3 presents the results, which are further interpreted in Section 4. Main findings are stated in Section 5.

2. Materials and Methods

2.1. Database

Initial database consisted of 61 paroxysmal AF patients without any previous CA sessions. Twenty-one patients were discarded due to extremely low amplitude or presence of noise and artifacts in the extreme channels of the CS recordings, probably due to dynamical change of mitral annulus anatomy during SR and vigorous myocardial contraction causing the movement of the CS cathether. The final database consisted of the remaining 40 patients.

Recordings from a standard 12-lead electrocardiogram (ECG) and a decapolar CS catheter with sampling frequency at 1 kHz were acquired by a Labsystem™ PRO EP recording system (Boston Scientific, Marlborough, MA, USA). Five-minute continuous segments before RFCA initiation (step **B**), after LPVI (step **L**) and after RPVI, which coincides with the end of the RFCA procedure (step **R**), were chosen. The evolution of each ablation step is illustrated in Figure 1a. Step **B** corresponds to 0% of the procedure, step **L** corresponds to 100% of LPVI, 0% of RPVI and 50% of the total ablation procedure, while step **R** corresponds to 100% of the overall procedure. The effect of ablation of each PV side is shown in Figure 1b. A statistically significant difference in features between steps **B** and **L** indicates that LPVI is critical in atrial substrate modification, while a difference between steps **L** and **R** or between **B** and **R** but without the difference between steps **B** and **L** being statistically significant, proves a significant effect of RPVI. Surface analysis was limited to lead II, as P-waves are more prominent in this lead [56].

For the study of the reliability of CS channels in preserving the AF dynamics, final database consisted of a total of 58 step **B** or **R** recordings from the 40 patients of the database. CS catheter consisted of the following channels of bipolar signals: distal (D), mid-distal (MD), medial (M), mid-proximal (MP) and proximal (P), with D channel being the closest to LA and P channel to the RA. Firstly, a multichannel comparison allowed us to define the channels that can record the AF dynamics to the most reliable degree. Selection of the analysis channel for the CS study was then performed among the most robust channels, with unique favorable criteria the high signal amplitude and low baseline fluctuation. Specific attention was paid so that the same channel would be employed for all CA steps for the same patient.

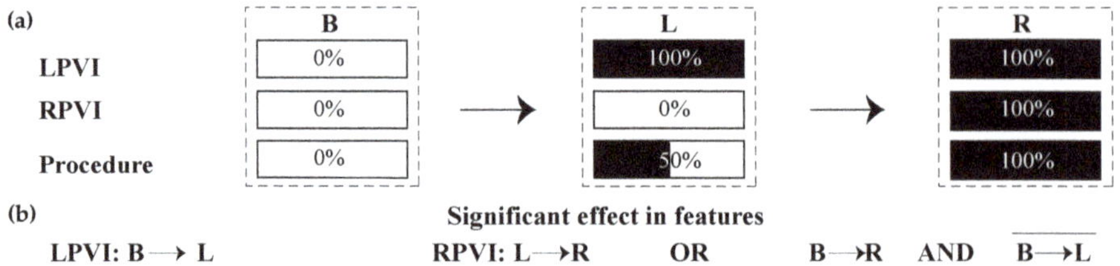

Figure 1. (**a**) Steps of CA procedure for which recordings were extracted and analyzed. In step **B**, no ablation has been performed yet (0%). In step **L**, LPVI has been completed (100%) and hence, we are in the middle of the procedure (50%). Step **R** corresponds to RPVI and to the end of the procedure. Therefore, each step is completed (100%). (**b**) Conditions in order for LPVI or RPVI to have a significant effect on the features under analysis. LPVI: left pulmonary vein isolation; RPVI: right pulmonary vein isolation.

Regarding the CA procedure, all patients underwent circumferential RFCA of PVs, guided by 3-D electroanatomical mapping during SR. RFCA was initiated by performing a crown surrounding left PVs (step **L**), followed by a crown surround right PVs (step **R**). Non-inducibility of AF was confirmed by pacing in all patients and was the endpoint of the procedure.

2.2. Preprocessing

For ECG recordings, powerline interference and high frequency muscle noise were removed by a wavelet-based denoising method [57] followed by a bidirectional low-pass filter with cut-off requency at 70 Hz [58]. Baseline wander was also removed [58].

EGM denoising and mean removal were the first preprocessing steps for CS recordings analysis [59]. Although presence of ventricular activity is not dominant in atrial bipolar signals, far-field activity in line with the R-peak of the ECG recordings has been observed in some cases. Removal of ventricular activity was performed with an adaptive cancellation method [60].

Next, ectopic beats correction was performed. Ectopic beats are premature atrial or ventricular contractions that affect the HRV. In our analysis, ectopic beats in ECGs, if present, did not exceed 4% of total beats. Their correction included the detection and cancellation of the premature complexes and their replacement by a new beat via linear interpolation [61]. Among various ectopic replacement methods, linear interpolation was chosen due to its better performance for time-domain HRV features [62].

Finally, detection and delineation of atrial activations was carried out. P-waves were firstly detected by an adaptive search window prior to the R-peak [63] and then delineated [64]. Local activation waves (LAWs) of CS were detected with an algorithm based on an alternative Botteron's technique [65]. Delineation was performed by firstly smoothing the LAW with a five-point moving average filter [66]. Delineation of both ECG and intracardiac recordings was visually inspected and corrected, if needed, by an expert.

2.3. Main Analysis

Once preprocessed, the duration, amplitude, root mean square (RMS) value, area, number of deflections and inflections (NODI) and slope rate were calculated for P-waves and LAWs, as shown in Figure 2. Final values of these features were calculated by signal-averaging. A brief description of these characteristics is provided as follows. Further details are described elsewhere [66].

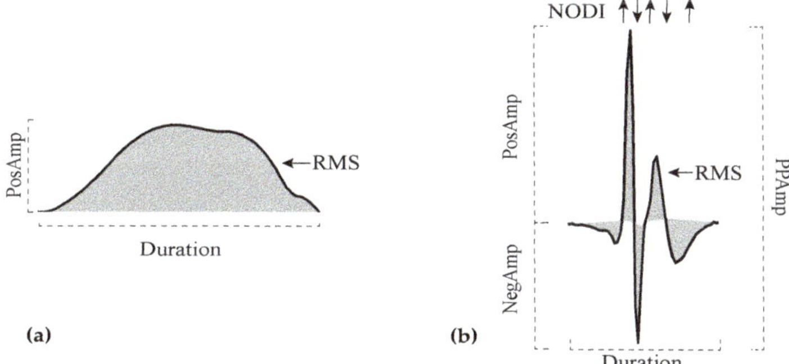

Figure 2. (**a**): Duration, amplitude, RMS and area (shaded) for a P-wave. (**b**): Duration, amplitude, RMS, area (shaded) and NODI for a LAW. Upward arrows represent an inflection, while downward arrows a deflection. In the figure, LAW has 3 major inflections and 2 major deflections. RMS: root mean square; NODI: number of deflections and inflections; LAWs: local activation waves.

1. *Duration:* Distance between the onset and offset of each activation.
2. *Amplitude:* Amplitude of positive and negative maximum of each activation were considered as positive (*PosAmp*) and negative amplitude (*NegAmp*), respectively. Peak-to-peak amplitude (*PPAmp*) was the distance between positive and negative maximum points. As P-waves are positive in lead II, only maximum amplitude was calculated for ECG analysis.
3. *RMS:* Let X_n be a time-series, so that $X_n = \{x_1, x_2, \ldots, x_n\}$. RMS value is the quadratic mean of the function that defines the time-series. In our case, this function is defined by either the P-wave or LAW waveform.
4. *Area:* Area is calculated as the integration of the signal over the time interval. Trapezoidal method allows this integration, by splitting each signal into smaller and easier to calculate trapezoids. Final area is defined by the cumulative sum of these trapezoids. As LAWs contain both positive (*PosAr*) and negative (*NegAr*) parts, this method was separately applied to each one of them.
5. *NODI:* Deflections and inflections were calculated from the points that cross two auxiliary baselines, at $\pm 25\%$ of the signal amplitude. This metric was only calculated for LAWs, as P-waves do not show multiple major deflections and inflections.
6. *Slope rate:* The rhythm of increasing or decreasing slope was calculated at sample points equal to $i\%$ of the activation duration, with $i = 5$, 10, 20. Slope rate at the maximum point was also computed. The equation calculating these slope rates was the following:

$$S_i = \frac{Amp(t_i) - Amp(t_{onset})}{t_i - t_{onset}}, \tag{1}$$

where $Amp(t_i)$ is the amplitude at the $i\%$ of the activation duration, $Amp(t_{onset})$ is the amplitude at the onset of the activation, t_i is the sample point at the $i\%$ of the activation duration and t_{onset} is the sample point corresponding to the onset of the activation.

Afterwards, features calculated across each recording were analyzed: morphology variability (MV), dispersion and time-domain HRV features.

1. *MV:* A reference signal was firstly created by the 20 most similar activations of the channel under analysis and then correlated with each and every activation, using an adaptive signed correlation index (ASCI) with 12% tolerance [67]. MV was then defined as the percentage of signals that correlated <95% with the reference signal.
2. *Dispersion:* Traditionally, for the calculation of dispersion, more than one ECG lead is employed and the difference between maximum and minimum activation duration across channels is computed. Alternatively, in our case, lead II was just extracted

and P-wave dispersion analysis was performed in this channel because atrial activity presents the highest amplitude. Dispersion was then defined as the difference between the 25th and 75th percentiles of the atrial activations duration of each recording. This way, the effect of signal delineation accuracy is minimized and an extremely long or short activation caused by various factors will not affect significantly the results [68]. Dispersion of EGM recordings was calculated in the same way.

3. *Time-domain HRV features:* HRV analysis is normally performed on R-R intervals, thus describing ventricular response. As in the present work we are focused on atrial analysis, we modified the techniques by substituting R-R peaks by P-wave to P-wave for ECG and LAW to LAW for EGM recordings. As these features describe the atrial response, thus neglecting the effect of the atrioventricular node and other cardiac structures, they will be referred in the remainder of this document as atrial rate variability (ARV) features. Standard deviation of normal-to-normal beat interval (*SDNN*), variance of normal to normal beat interval (*VARNN*) and RMS of successive interbeat differences (*RMSSD*) were calculated for each recording.

Heart Rate Adjustment

Time-domain features of P-wave analysis are affected by variable HR [69]. More specifically, as HR increases, intervals between fiducial points of ECGs shorten. Therefore, HR adjustment is proposed in order to moderate this effect. For this purpose, additionally to the original analysis, a simple HR-adjustment factor is performed. As sampling frequency is 1 kHz, a 60 beat-per-minute recording would show one activation every 1000 sample points. However, as HR is diverse and often deviant from these values, the adjustment factor for the ith activation was set as

$$adj(i) = \frac{1000}{IBI_i},\tag{2}$$

where IBI_i is the interbeat interval between the ith and the $(i-1)^{\text{th}}$ activations. Duration and area were normalized by this factor, while slope rates were inversely scaled by it. HR-adjusted values will be shown as **HRA**(y), where $y = duration, area$ or S_i.

2.4. Statistical Analysis

Normality and homoscedasticity were tested with Saphiro-Wilk and Levene tests, respectively [70,71]. According to the results, non-parametric tests were employed for the comparison between populations.

Reliability analysis of the CS channels with respect to AF dynamics was performed on *Duration, Amplitude, RMS, Area* and *NODI*. In the first place, a multichannel comparison via a Kruskal-Wallis (KW) test was employed [72]. Comparison in pairs of two channels was performed with a Mann-Whitney U-test (MWU) [73] with Bonferroni correction. Median values were also calculated at each channel and any significant differences between each one and the remaining channels was explored by as well using a MWU with Bonferroni correction. Afterwards, a reference signal for each channel was calculated as described from MV analysis in Section 2.3. Then, the correlation between the morphology of each channel's reference signal in pairs of two was calculated for each recording, using an ASCI with 12% tolerance.

For P-waves, LAWs and ARV analysis, comparison between ablation steps is performed with KW and post-hoc tests to define which step is crucial are performed with MWU with Bonferroni correction and median and interquartile range calculations. Analysis is performed for P-waves and CS LAWs separately. Percentage of variation (POV) of features was specified for each recording and between two ablation steps as

$$POV(r_i) = (\frac{V_2}{V_1} - 1) \times 100 \; [\%],\tag{3}$$

where r_i is the recording of the *i*th patient, V_2 is the value of each feature at the posterior step and V_1 is the value of the same feature at the prior step. How POV was modified between ablation steps **B-L** and **L-R** for both P-waves and LAWs was tested with MWU. Finally, HR was measured at each ablation step and compared among all steps with KW.

3. Results

3.1. Analysis of CS Features between Channels

Table 1 shows the results of the multichannel comparison as well as the comparison in pairs of channels, for the selected features. *Amplitude* and *Area* show different values among the CS catheter channels. Paired analysis showed that differences are mostly located between D and M or D and MP channels. A trend has also been observed between values of MD and M channels. Due to Bonferroni correction, α is 0.005, as 10 paired comparisons have been performed. Comparison between D-MD, MD-P and M-MP channels are not illustrated, as they did not show any statistically significant differences ($p > 0.03, 0.38$ and 0.35, respectively).

Table 1. Multichannel comparison (KW) and paired comparison (MWU) for the features defined in CS channels. Statistically significant results are shown with an asterisk (*). In MWU analysis, $\alpha = 0.005$. D: distal; MD: mid-distal; M: medial; MP: mid-proximal; P: proximal.

| | **KW** | **MWU** | | | | | | |
		D-M	**D-MP**	**D-P**	**MD-M**	**MD-MP**	**M-P**	**MP-P**
Duration	0.2136	0.0188	0.0893	0.4355	0.3502	0.8927	0.2232	0.7001
PosAmp	<0.0001 *	<0.0001 *	<0.0001 *	0.0140	0.0053	0.0038 *	0.0355	0.0502
NegAmp	0.0005 *	0.0002 *	0.0016 *	0.4162	0.0087	0.0587	0.0080	0.0428
PPAmp	0.0001 *	<0.0001 *	0.0002 *	0.1476	0.0087	0.0182	0.0112	0.0491
RMS	0.0003 *	0.0001 *	0.0007 *	0.1644	0.0053	0.0409	0.0182	0.0950
PosAr	0.0008 *	0.0003 *	0.0020 *	0.1826	0.0080	0.0491	0.0182	0.0950
NegAr	0.0024 *	0.0014 *	0.0025 *	0.3384	0.0119	0.0347	0.0207	0.0758
Deflections	0.7925	0.6217	0.2458	0.3258	0.9749	0.5681	0.6126	0.7430
Inflections	0.8045	0.6324	0.7477	0.3711	0.9840	0.4349	0.7831	0.2354

Figure 3 shows the median values obtained for each analyzed feature from the CS. As can be seen, the distal channel showed the longest LAWs *Duration* and lowest LAWs *Amplitude* and *Area* values for most of the features. Medial channel contained the shortest LAWs *Duration* and highest LAWs *Amplitude* values. *Area* was higher in mid-proximal channel, followed by medial channel. Mid-proximal channel showed overall similar behavior as medial channel, with rather high *Amplitude* and short LAWs *Duration* values.

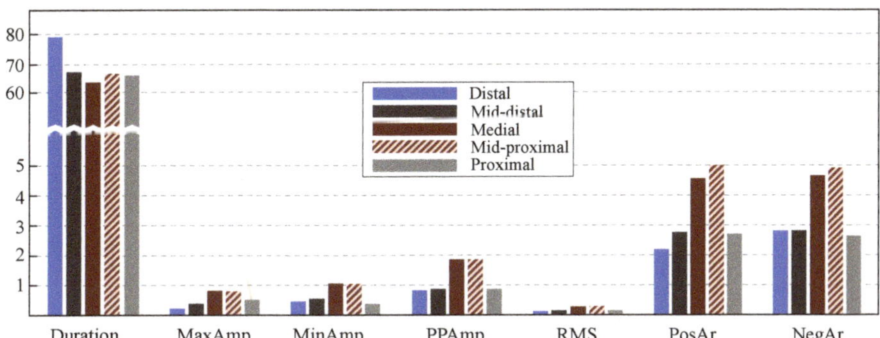

Figure 3. Bar graph for the median values of the analyzed features at each one of the CS channels. Note the break in the vertical scale for the feature *Duration*.

The analysis of channels that varied significantly with respect to the others at each feature is shown in Table 2. Statistical comparison allows the detection of the channels that show the most and least deviant values and can corroborate the results shown in Figure 3.

Table 2. One-vs.-all analysis of defined features for each one of CS channels. Asterisks (*) indicate statistically significant values. Due to multiple comparison, α has been modified to 0.01.

Features	Distal	Mid-Distal	Medial	Mid-Proximal	Proximal
Duration	0.0472	0.8617	0.0837	0.6755	0.7345
PosAmp	<0.0001 *	0.1688	0.0021 *	0.0061 *	0.8176
NegAmp	0.0038 *	0.4720	0.0019 *	0.0446	0.1317
PPAmp	0.0008 *	0.2669	0.0017 *	0.0204	0.3261
RMS	0.0014 *	0.2762	0.0018 *	0.0394	0.3717
PosAr	0.0022 *	0.3799	0.0027 *	0.0610	0.3555
NegAr	0.0103	0.2427	0.0049 *	0.0526	0.3113
Deflections	0.3181	0.8899	0.8099	0.3745	0.6235
Inflections	0.6108	0.7069	0.7779	0.3422	0.4243

None of the channels showed a statistically different LAWs *Duration* comparing to the remaining ones. Nevertheless, a trend was observed for distal and medial channels. Distal and medial channels showed additionally statistically different values with respect to the remaining channels regarding LAWs *Amplitude* and *Area* features. Mid-proximal channel also showed statistically significant difference than the other channels for positive *Amplitude* and a trend for the remaining *Amplitude* and *Area* features. Combining the aforementioned observations with the median values of the LAWs features for each CS channel presented in Figure 3, we can conclude that *Amplitude* and *Area* are statistically smaller in distal channel and higher in medial channel of the catheter, while duration tends to be longer in distal and shorter in medial channels. Mid-proximal channel shows a trend for high *Amplitude* and *Area* values as well.

Finally, how LAWs morphology of each channel correlated with the morphology of LAWs at each of the remaining channels can be appreciated from Figure 4. Channels of proximal area (MP, P) show higher correlations between their LAWs morphology compared to correlations from distal area (D, MD). Additionally, medial channel showed stronger correlation with proximal (93–95%) than distal area (86–91%). Although all adjacent channels showed relatively strong correlations, the highest values were observed in proximal area.

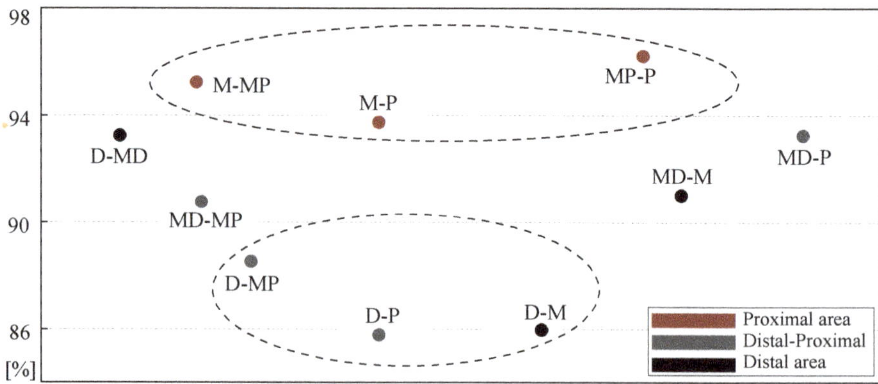

Figure 4. Correlations, as percentage, between the LAWs morphology in various CS catheter channels. The upper ellipse area contains the strongest and the bottom ellipse area the weakest correlations. Stronger correlations are found in proximal area, while moderate or weaker correlations are observed in distal area or between channels that are spatially far away (distal-proximal area).

3.2. Analysis of Features from P-waves and LAWs

The HR measurements did not reveal any statistical difference between HR at different ablation steps. However, a decrease in HR was observed in step **L**, as shown in Table 3.

Table 3. Heart-rate at each ablation step and comparison between three steps (KW). As result indicated a non-significant comparison, no MWU has been performed. KW: Kruskal-Wallis; MWU: Mann-Whitney U-test; B: before CA; L: after LPVI; R: after RPVI.

	B	L	R
Median (iqr)	57.2 (17.0)	55.0 (12.0)	58.6 (13.4)
KW		0.7713	

Table 4 shows the median and interquartile range for the calculated features in P-waves at each ablation step. Multiple comparison among ablation steps is then shown in the fifth column (KW) and comparison of each feature between ablation steps in pairs of two using Bonferroni correction can be observed in the last three columns (MWU). *Duration* varied statistically among channels. When analysis in pairs was conducted, this variation was detected between steps **B-L** and **B-R**. As **L-R** comparison did not show any significant variation, the **B-R** significance is probably due to the **B-L** significance. Hence, step **L** is considered the critical step for the reduction in *Duration*, which falls from 120 ms in the beginning of the procedure to 104 ms, almost the final value observed after the end of CA of PVs.

Table 4. Median (interquartile) values for each feature and results for KW and MWU tests for P-waves. Statistically significant results are shown in (*). Due to Bonferroni correction, threshold for MWU (last three columns) is $\alpha = 0.0167$.

	Median			KW	MWU		
Features	B	L	R		B-L	B-R	L-R
Duration [ms]	120.0 (12.00)	104.0 (13.00)	106.5 (21.00)	0.003 *	0.001 *	0.009 *	0.558
PosAmp [mV]	0.428 (0.303)	0.354 (0.290)	0.374 (0.232)	0.084	0.055	0.097	0.319
PPAmp [mV]	0.431 (0.303)	0.356 (0.290)	0.374 (0.232)	0.084	0.056	0.097	0.319
RMS [mV]	0.263 (0.179)	0.214 (0.179)	0.230 (0.232)	0.144	0.103	0.150	0.275
PosAr [mV$\times$ms]	24.63 (12.94)	16.57 (14.62)	20.39 (9.98)	0.141	0.103	0.103	0.438
S_5 [mV/ms]	0.005 (0.002)	0.007 (0.004)	0.006 (0.002)	0.162	0.110	0.235	0.211
S_{10} [mV/ms]	0.007 (0.002)	0.008 (0.004)	0.007 (0.003)	0.178	0.117	0.420	0.150
S_{20} [mV/ms]	0.010 (0.005)	0.011 (0.003)	0.009 (0.005)	0.336	0.384	0.693	0.133
S_{max} [mV/ms]	0.010 (0.004)	0.009 (0.004)	0.008 (0.004)	0.823	0.987	0.602	0.602
HRA(*Duration*)	119.5 (57.39)	106.9 (26.04)	101.0 (36.91)	0.159	0.141	0.079	0.740
HRA(*PosAr*)	26.10 (16.94)	19.40 (14.16)	22.12 (14.01)	0.144	0.085	0.110	0.716
HRA(S_5)	0.004 (0.004)	0.006 (0.006)	0.006 (0.003)	0.367	0.261	0.235	0.537
HRA(S_{10})	0.006 (0.003)	0.007 (0.007)	0.007 (0.004)	0.441	0.248	0.402	0.558
HRA(S_{20})	0.010 (0.005)	0.010 (0.008)	0.009 (0.006)	0.801	0.693	0.837	0.517
HRA(S_{max})	0.009 (0.007)	0.008 (0.006)	0.008 (0.006)	0.994	0.962	1.000	0.912
MV	0.605 (0.329)	0.753 (0.335)	0.675 (0.467)	0.476	0.189	0.624	0.646
Dispersion [ms]	12.00 (4.000)	11.00 (7.000)	10.00 (4.000)	0.310	0.208	0.176	0.949
SDNN	94.25 (55.32)	99.91 (71.96)	84.28 (62.10)	0.136	0.133	0.862	0.060
VARNN	8.8×10^3 (1.07×10^4)	9.9×10^3 (1.77×10^4)	7.1×10^3 (1.21×10^4)	0.136	0.133	0.862	0.060
RMSSD	95.44 (59.68)	126.79 (95.29)	92.51 (61.73)	0.136	0.052	0.962	0.069

In the same context, amplitude showed a downward trend after step **L**. Despite the key role of step **L** in the modification of *Duration*, when HR adjustment was performed, this step did not show any effect in **HRA**(*Duration*) and neither did step **R**. However, modification of **HRA**(*Duration*) showed a trend when values in the beginning and the end of the procedure were measured (**B-R** comparison), possibly due to a cumulative effect of CA in AF substrate which can be better appreciated quantitatively when isolation is totally performed. Although *Area* modification was not originally found to be statistically significant at each of the ablation steps, HR adjustment revealed for it a trend in **B-L** comparison, falling from 26.10 mV×ms to 19.40 mV × ms. Finally, it can also be observed that measurements after step **L** showed, in a non-significant level, lower *Amplitude* and *Area* values and higher ARV, *MV* values and *Slope rate* values than steps **B** and **R**.

Regarding LAWs, comparison of each feature among all channels revealed a significant variation of *RMSSD*. Apart from this observation, none of the features varied statistically at any of the ablation steps. However, some trends have been observed. The respective results are shown in Table 5. LAWs showed a trend for shortening between steps **B-R**, as shown by both *Duration* and **HRA**(*Duration*) results. *MV* showed a trend for amplification after step **L** (**B-L** comparison) which was almost statistically significant. All ARV features employed in this study showed an increasing trend after step **L** (**B-L** comparison) and a decreasing trend after step **R** (**L-R** comparison). Note that threshold for MWU is $\alpha = 0.0167$ due to Bonferroni correction.

Table 5. Median (interquartile) values for each feature and results for KW and MWU tests for CS LAWs. Statistically significant results are shown in (*). Due to Bonferroni correction, threshold for MWU (last three columns) is $\alpha = 0.0167$. As highest peak is often found in negative amplitude in LAWs, slope rate in **HRA**(S_{max}) is negative.

Features	Median B	Median L	Median R	KW	MWU B-L	MWU B-R	MWU L-R
Duration [ms]	100.5 (14.00)	97.50 (18.00)	90.00 (23.00)	0.108	0.241	0.055	0.217
PosAmp [mV]	0.492 (0.987)	0.509 (0.893)	0.641 (0.775)	0.835	0.646	0.887	0.602
NegAmp [mV]	−0.831 (1.572)	−0.779 (0.772)	−0.915 (0.486)	0.892	0.740	0.912	0.646
PPAmp [mV]	1.361 (2.624)	1.382 (1.495)	1.570 (1.462)	0.942	0.740	0.937	0.837
RMS [mV]	0.150 (0.345)	0.151 (0.195)	0.181 (0.218)	0.847	0.740	0.912	0.558
PosAr [mV×ms]	4.407 (5.175)	3.718 (5.160)	3.985 (4.836)	0.896	0.670	0.837	0.788
NegAr [mV×ms]	4.085 (5.485)	3.818 (4.045)	3.992 (4.655)	0.864	0.624	0.937	0.693
Deflections	3.000 (1.000)	3.000 (1.000)	3.000 (1.000)	0.916	0.682	0.889	0.828
Inflections	3.000 (1.000)	3.000 (1.000)	2.500 (1.000)	0.915	0.878	0.810	0.695
S_5 [mV/ms]	3.2×10^{-4} (2.1×10^{-4})	3.9×10^{-4} (3.4×10^{-4})	3.8×10^{-4} (5.0×10^{-4})	0.732	0.624	0.837	0.438
S_{10} [mV/ms]	4.3×10^{-4} (6.0×10^{-4})	4.4×10^{-4} (5.2×10^{-4})	4.9×10^{-4} $(4.8 \times 10^{-4}$	0.767	0.764	0.558	0.558
S_{20} [mV/ms]	3.8×10^{-4} (0.003)	4.7×10^{-4} (0.001)	5.3×10^{-4} (0.001)	0.932	0.887	0.962	0.669
S_{max} [mV/ms]	−0.018 (0.058)	−0.019 (0.027)	−0.021 (0.044)	0.996	0.937	0.987	0.987
HRA(*Duration*)	104.8 (32.83)	102.1 (23.50)	91.25 (33.19)	0.159	0.669	0.060	0.200
HRA(*PosAr*)	4.446 (6.861)	3.503 (4.345)	3.832 (4.737)	0.882	0.669	0.693	0.937
HRA(S_5)	3.6×10^{-4} (1.9×10^{-4})	4.1×10^{-4} (3.9×10^{-4})	3.4×10^{-4} (4.1×10^{-4})	0.753	0.558	0.837	0.517
HRA(S_{10})	4.4×10^{-4} (0.001)	4.2×10^{-4} (3.6×10^{-4})	4.7×10^{-4} (0.001)	0.771	0.912	0.558	0.537
HRA(S_{20})	4.1×10^{-4} (0.003)	3.9×10^{-4} (0.001)	5.1×10^{-4} (0.001)	0.836	0.710	0.812	0.580
HRA(S_{max})	−0.015 (0.049)	−0.016 (0.030)	−0.019 (0.047)	0.967	0.887	0.788	0.912
MV	0.028 (0.048)	0.100 (0.103)	0.067 (0.351)	0.056	0.018	0.113	0.692
Dispersion [ms]	2.500 (3.000)	2.000 (4.000)	2.000 (6.000)	0.676	0.461	0.923	0.451
SDNN	74.18 (57.56)	96.51 (94.74)	61.75 (73.48)	0.056	0.048	0.912	0.041
VARNN	5.5×10^3 (7.9×10^3)	9.4×10^3 (2.0×10^4)	3.4×10^3 (1.1×10^4)	0.056	0.048	0.912	0.041
RMSSD	98.87 (86.99)	127.4 (119.8)	90.40 (74.41)	0.049 *	0.026	0.764	0.064

POV of all features between each two ablation steps are illustrated in Table 6 for both P-waves and LAWs. Comparison of POV that corresponds to successive step transitions **B-L** and **L-R** can also be seen in the last two columns. In general, POV in P-waves seem to be more prominent than respective POV values in LAWs, especially in the **B-L** comparison. *Duration* in P-waves got reduced by −13.30% after step **L**, while in LAWs by −5.49%. Step **R** did not additionally modify at average the P-waves at all (0.00% POV in **L-R** comparison). For LAWs, step R did reduce LAW *Duration* by an additional −1.93%. Comparison between **B-L** and **L-R** alterations in P-wave *Duration* presented a significant difference ($p < 0.0001$), which was not observed in LAWs, indicating a different effect of LPVI and RPVI in P-wave but not in CS LAWs *Duration* alteration.

Table 6. POV for between every two ablation steps for P-waves and LAWs and comparison between POV of successive step transitions **B-L** and **L-R**. Statistically significant results are shown in (*). POV: percentage of variation.

| Features | B-L [%] | | L-R [%] | | B-R [%] | | MWU (BL-LR) | |
	P-Waves	LAWs	P-Waves	LAWs	P-Waves	LAWs	P-Waves	LAWs
Duration	−13.30	−5.49	0.00	−1.93	−11.01	−7.46	<0.0001	0.6576
PosAmp	−16.65	−3.58	3.86	0.52	−11.25	−6.51	0.0556	0.6464
PPAmp	−16.04	−7.79	3.80	−2.15	−11.52	−5.09	0.0556	0.7397
RMS	−18.51	0.82	7.51	19.91	−12.39	20.89	0.1032	0.7397
PosAr	−25.68	−5.23	6.50	−1.74	−20.72	−4.06	0.1032	0.6693
S_5	30.95	22.47	−13.75	−1.82	12.95	20.24	0.1101	0.6239
S_{10}	15.75	1.34	−11.23	11.74	−0.73	13.24	0.1173	0.7637
S_{20}	12.84	25.89	−18.35	12.20	−7.87	41.24	0.3843	0.8868
S_{max}	−5.40	4.37	−6.96	7.75	−11.99	12.46	0.9874	0.9370
HRA(*Duration*)	−13.73	−3.89	1.60	−4.46	−14.49	−9.84	0.1412	0.6693
HRA(*PosAr*)	−24.54	−2.90	6.60	−3.86	−19.65	−19.94	0.0847	0.6693
HRA(S_5)	58.69	14.18	−9.34	−16.49	43.87	−4.65	0.2614	0.5583
HRA(S_{10})	13.41	−4.10	−4.38	11.80	8.45	7.21	0.2482	0.9118
HRA(S_{20})	3.35	−5.64	−6.44	32.58	−3.30	25.10	0.6925	0.7160
HRA(S_{max})	−6.60	4.11	−1.21	17.70	−7.74	22.53	0.9621	0.8868
MV	15.00	144.9	−6.11	−5.93	−0.42	172.1	0.1892	0.0176 *
Dispersion	−22.42	0.00	22.22	80.00	−9.55	0.00	0.2084	0.4613
SDNN	28.39	79.80	−24.27	−33.94	0.28	0.92	0.1249	0.0480 *
VARNN	64.86	225.9	−42.64	−55.91	0.57	1.93	0.1249	0.0480 *
RMSSD	45.15	73.12	−28.30	−36.30	0.26	5.43	0.0445 *	0.0257 *

HR-adjustment did not have a different effect in **B-L** comparison of P-waves with respect to the same comparison for non-normalized *Duration* variation (−13.73% vs. −13.30%, respectively). However, HR-adjustment revealed a slightly higher, albeit not statistically significant, effect of step **R** in **HRA**(*Duration*), showing an incrementation of +1.60%. For LAWs, **HRA**(*Duration*) slightly mitigated the effect of step **L** (−3.89% for HR-adjustment) and potentiated the effect of step **R** (−4.46%). These results were not statistically significant either. *Amplitude* and **HRA**(*PosAr*) values showed the same kind of variations for P-waves and LAWs, with comparison between **B-L** and **L-R** transitions showing a trend for P-waves and low statistical power for LAWs.

For the remaining features, *MV* showed a rather strong magnification after step **L** by +144.90% in LAWs, whereas the corresponding step in P-waves showed only a +15.00% of *MV* magnification. After step **R**, *MV* dropped by −5.93% in LAWs and by −6.11% in P-waves. Due to very intense variations in *MV* of LAWs, POV evolution between transitions from **B-L** and **L-R** varied significantly ($p = 0.0176$). As already observed in Tables 4 and 5, ARV features showed a notable incrementation after step **L** in both P-waves and LAWs. This is even more prominent in POV analysis, where ARV got increased by up to +225.90% in LAWs and up to +64.86% in P-waves (in *VARNN* in both cases). However, ARV after the end of the procedure (step **R**) got decreased by up to −55.91% in LAWs and −42.64% in

P-waves (in *VARNN* in both cases). These very high variations in LAWs analysis led the **B-L** and **L-R** comparison to be statistically significant for all ARV features. As P-waves POV analysis also showed considerable variations but to a lesser extent, only POV in *RMSSD* varied statistically between **B-L** and **L-R**. Figure 5 then shows the POV box and whisker plots for the features that showed any kind of significant alteration or a trend, so that alterations in POV can be better illustrated. From $y-$axis can be additionally observed that POV in LAWs shows more scattered values that span along a wider range than P-waves analysis. This is noticeable at most of the boxplot pairs of Figure 5, but can be especially seen in the subplot of *MV*, where POV in P-waves is in the range of 0 to 1200% while in LAWs in the range of 0 to 12,000%.

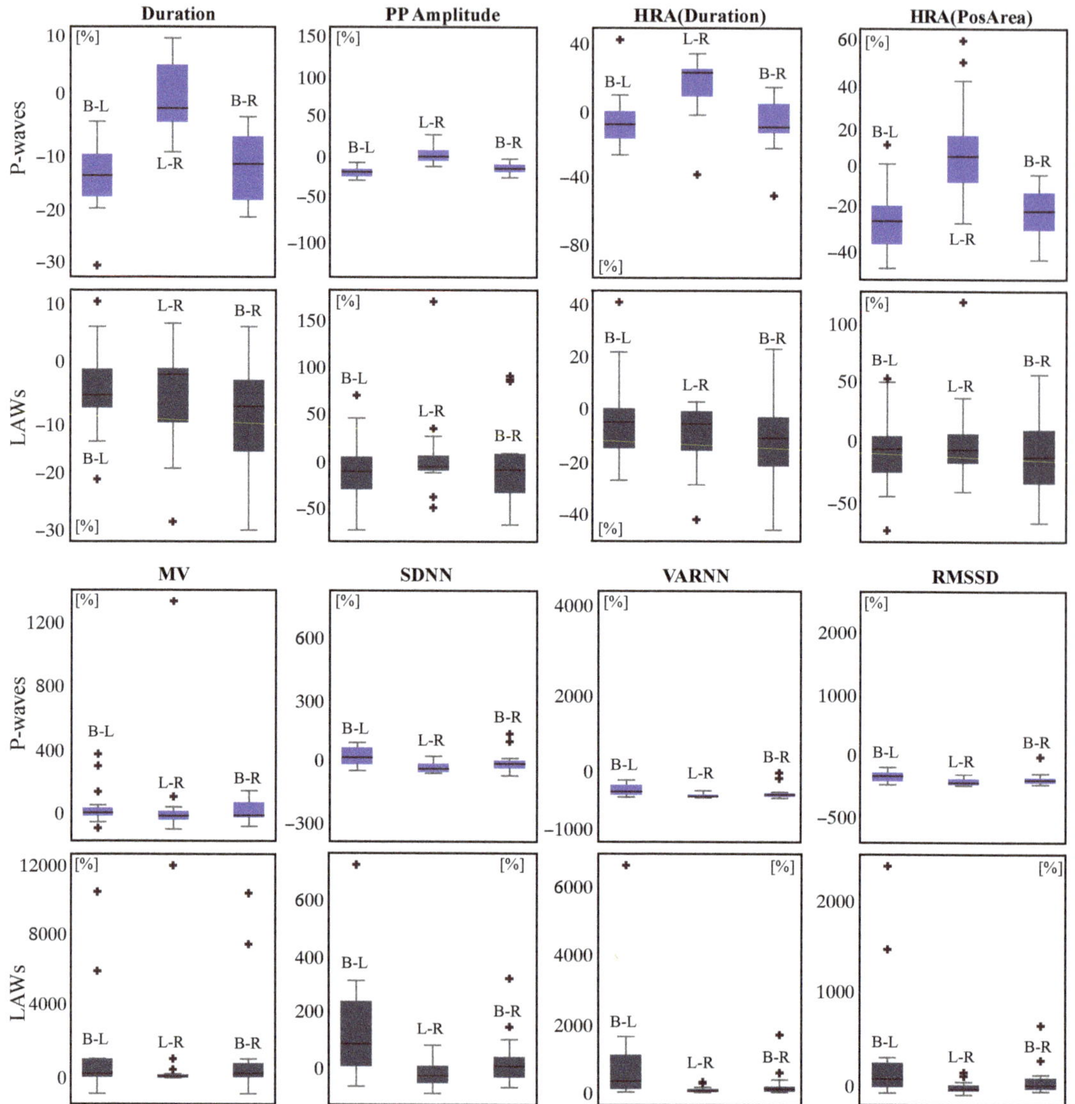

Figure 5. Representation of significant alterations illustrated as boxplots of POV for P-waves features (each **top**, blue) and LAWs features (each **bottom**, gray) between the defined ablation steps for selected features. Beware of the different vertical scale in each case for the same feature.

4. Discussion

This multi-approach study had three main objectives. First of all, to define the CS channels that can record with the highest precision and robustness the AF dynamics during SR. The analysis revealed the existence of variability among CS channels especially in *Duration* and *Amplitude* features. Differences were mostly found betweeen distal and medial and between distal and mid-proximal, with a trend between mid-distal and medial channels. A combined interpretation of the analysis of medians and the one-vs-all analysis indicates that distal channel showed the longest *Duration*, whilst the shortest *Duration* has been recorded by the medial channel. Regarding *Amplitude* and *Area* values, these were smaller in distal channel and larger in medial and mid-proximal channels. Proximal area showed the strongest morphological correlations between its channels. On the contrary, correlations in distal area or between distal and proximal channels were weaker. Hence, being the least susceptible channels to exogenous factors during SR, medial and mid-proximal channels are recommended while distal and mid-distal channels are not suggested.

Various studies corroborate these conclusions. CS EGM fractionation analysis was related with AF recurrence during SR in proximal and medial but not in distal channel in a recent study [47]. Fractionation of proximal CS EGMs indicated AF patients in another study employing recordings during AF [74]. Another work found that AF cycle length analysis in distal channel failed to predict AF termination after CA. In the same study, only mid-proximal channel predicted freedom from AF recurrence [75].

The second objective of the present study was to investigate if CS recordings can describe adequately the substrate modification due to CA, as observed by P-wave analysis. Parallel P-waves and LAWs analysis has been conducted for this purpose. The former represent the entire atria while the latter provide very specific yet crucial information on CS function. Variations were observed to a higher degree in most of the features in P-waves than LAWs. *MV* and ARV features were modified to an exceptionally higher extent in LAWs than P-waves. As CS recordings are closer to the tissue under ablation than surface ECG recordings, variability caused by RF energy deliverance may be illustrated with higher precision by LAWs [38,39]. Variation was more consistent in P-waves than LAWs, while the latter showed higher *Dispersion* in values across all features.

The last but not least purpose of this work was the evaluation of additional recordings acquired during CA in order to understand the role that the ablation of each PV side plays to the modification of the studied features and, as a consequence, to the atrial substrate alteration. A significant P-wave shortening was observed after CA of both PV sides, in line with a plethora of previous studies mainly attributed to fibrotic areas causing conduction delays [20–23]. Interestingly enough, this reduction was observed right after LPVI, with RPVI not showing any additional effect to this feature. *Duration* before CA was 120 ms, dropping down to 104 ms after LPVI and showing a minor increase after RPVI, to 106.5 ms. P-wave *Amplitude* also tended to show a lower value after LPVI which was slightly increased after RPVI, but remained overall smaller than the pre-ablative measurements.

HRV attenuation after CA is considered an indicator of CA success [38–40]. In line with previous studies, ARV in the present study showed a non-significant reduction after the end of the procedure. Nevertheless, recordings obtained after LPVI showed a trend for amplification of ARV values by up to +64.86%. Previous works studying the effect of RF energy in rabbits and students in lying position found that RF exposure can cause HRV incrementation and HR attenuation [76,77]. These findings explain the aforementioned results of the present study. Additionally, HR was indeed found to decrease after LPVI in the present work. Finally, HR-adjustment preserved the variation that was observed in *Duration* after LPVI, although losing statistical power. It also incremented the different effect that RPVI had on *Duration*, showing a slight non-statistical increase of +1.60% after RPVI. The reason for this slight incrementation is the HR acceleration after the end of CA procedure with respect to recordings after LPVI, where RF energy deliverance was still going on. Higher HR leads to generally narrower P-waves, the size of which is retrieved after HR-adjustment. An additional factor that may have a minor effect on this incrementation is a possible deviation of 1–2 ms in P-wave delineation precision, which due to its size (<+2.00% of duration values range) is considered acceptable.

Amplitude in P-waves showed a trend for reduction after LPVI. Although final *Amplitude* was non-statistically reduced with respect to the beginning of the procedure, as in previous studies [27], RPVI slightly but non-significantly increased *Amplitude* values. Contrastly, *Slope rate* was non-significantly increased after LPVI but decreased after RPVI in most of the studied time instances. It is highly possible that RF exposure also has had an effect on *Amplitude* and *Slope rate* features, explaining these variations.

Regarding LAWs, variations of most features show weaker statistical power and lower POV at each ablation step with respect to P-waves analysis. As mentioned afore, *MV* and ARV varied more prominently in LAWs than P-waves. *MV* showed a high incrementation in the order of +144.9% after LPVI and a decrease of −5.93% after RPVI. The reasons for these dramatic changes are not clear. Exposure to RF energy, not only affecting ARV but also *MV* may be an explanation. Recently a study found *MV* in lead V1 P-waves of paroxysmal AF undergoing CA of PVs to decrease after the procedure, as a sign of a successful ablation [78]. Apart from the fact that these results come from P-waves analysis, which in our case show a slight attenuation overall, *MV* in the present analysis was extracted by a template of the 20 most similar activations and not by considering all activations of one recording. ARV was also increased after LPVI in LAWs and to a higher extent than P-waves (up to +225.9%), possibly due to proximity to the tissue under ablation as explained previously.

To our knowledge, this is the first complete comparative study to perform simultaneous P-waves and LAWs analysis on recordings obtained not only before and after but also during CA of PVs. Recently, a relevant study calculated the organization of ECG recordings before, during and after CA of PVs and did not find any step to affect significantly the organization indices under calculation [79]. As many differences exist with respect to our study, a comparison would not be straightforward. In the first place, individuals studied were persistent AF patients while the present study employed exclusively paroxysmal AF patients. As persistent AF shows more complicated atrial substrate and often presents AF drivers outside of PVs, efficiency of CA of PVs is notably lower with respect to success rates in paroxysmal AF patients [19,80]. Secondly, the procedure consisted of CA of PVs, CA of CFAEs and linear CA of LA. Recordings during the procedure were the recordings after CA of PVs and before CA of CFAEs and linear CA of LA. Hence, the intermediate stage of their analyis would be the final stage of ours and no information is provided about the contribution of left or right PVs. Finally, features employed are different from the features employed in the present study.

The key aspects of CA of PVs for paroxysmal AF patients investigated in the present study improve significantly the understanding of the AF mechanisms during SR and contribute to the knowledge on how these mechanisms respond to each step of CA. Moreover, the CA procedure itself is reconsidered and the most reliable means to analyze CS EGMs are explored. Overall, a more detailed perspective of the CA procedure and the effect of RF exposure to atrial tissue is obtained.

5. Conclusions

LPVI is the critical part of CA of PVs for paroxysmal AF patients, altering significantly the P-wave duration. RF exposure tends to cause temporary ARV incrementation, which is reversed right after the end of the CA procedure. The effect of CA of PVs on CS is less straightforward and takes place to a lesser extent. Thus, other atrial structures may be more indicative of the ablation outcome and should be assessed as alternative references.

It should be noted, however, that ARV modifications regarding RF energy are more prominently observed in CS LAWs, possibly due to the vicinity with the tissue under RF exposure. Hence, the employment of CS recordings may be beneficial for the study of ARV alterations during and after CA of PVs.

Finally, studies interested in employing CS analysis are encouraged to extract and investigate medial or mid-proximal channels, as they were found to be the most robust, showing the highest coherence between LAWs morphologies. Distal and mid-distal channels, on the other hand, should be avoided as they were prone to variable morphology and less clear activations.

Author Contributions: Conceptualization, A.V., R.A. and J.J.R.; methodology, A.V., R.A. and J.J.R.; software, A.V.; validation, A.V., V.B.-G., L.F., J.M.-A., R.A. and J.J.R.; resources, V.B.-G., L.F., J.M.-A. and J.J.R.; data curation, A.V., V.B.-G., L.F. and J.M.-A.; original draft preparation, A.V.; review and editing, A.V., V.B.-G., L.F., J.M.-A., R.A. and J.J.R. All authors have read and agreed to the published version of the manuscript.

Funding: This research has received partial financial support from public grants DPI2017-83952-C3, PID2021-00X128525-IV0, PID2021-123804OB-I00 and TED2021-129996B-I00 of the Spanish Government 10.13039/501100011033 jointly with the European Regional Development Fund (EU), SB-PLY/17/180501/000411 from Junta de Comunidades de Castilla-La Mancha and AICO/2021/286 from Generalitat Valenciana.

Institutional Review Board Statement: The study was conducted according to the guidelines of the Declaration of Helsinki, complied with Law 14/2007, 3rd of July, on Biomedical Research and other Spanish regulations and was approved by the Ethical Review Board of Saint John's University Hospital (San Juan de Alicante, Alicante, Spain) with protocol code 21/046.

Informed Consent Statement: Written informed consent was granted from all the subjects participating in the present research. All acquired data were anonymized before processing.

Data Availability Statement: The data supporting reported results and presented in this study are available on request from the corresponding author.

Conflicts of Interest: The authors have no association with commercial entities that could be viewed as having an interest in the general area of the submitted manuscript. The funders had no role in the design of the study; in the collection, analyses, or interpretation of data; in the writing of the manuscript, or in the decision to publish the results.

References

1. Hindricks, G.; Potpara, T.; Dagres, N.; Arbelo, E.; Bax, J.J.; Blomstro, C.; Boriani, G.; Castella, M.; Dan, G.A.; Dilaveris, P.E.; et al. 2020 ESC Guidelines for the diagnosis and management of atrial fibrillation developed in collaboration with the European Association of Cardio-Thoracic Surgery (EACTS). *Eur. Heart J.* **2020**, *42*, 374–498. [CrossRef]
2. Haissaguerre, M.; Jaïs, P.; Shah, D.C.; Takahashi, A.; Hocini, M.; Quiniou, G.; Garrigue, S.; Le Mouroux, A.; Le Métayer, P.; Clémenty, J. Spontaneous initiation of atrial fibrillation by ectopic beats originating in the pulmonary veins. *N. Engl. J. Med.* **1998**, *339*, 659–666. [CrossRef] [PubMed]
3. Ioannidis, P.; Zografos, T.; Christoforatou, E.; Kouvelas, K.; Tsoumeleas, A.; Vassilopoulos, C. The Electrophysiology of Atrial Fibrillation: From Basic Mechanisms to Catheter Ablation. *Cardiol. Res. Pract.* **2021**, *2021*, 4109269. [CrossRef]
4. Lau, D.H.; Linz, D.; Schotten, U.; Mahajan, R.; Sanders, P.; Kalman, J.M. Pathophysiology of Paroxysmal and Persistent Atrial Fibrillation: Rotors, Foci and Fibrosis. *Heart Lung Circ.* **2017**, *26*, 887–893. [CrossRef] [PubMed]
5. Mouws, E.M.J.P.; Lanters, E.A.H.; Teuwen, C.P.; van der Does, L.J.M.E.; Kik, C.; Knops, P.; Bekkers, J.A.; Bogers, A.J.J.C.; de Groot, N.M.S. Epicardial Breakthrough Waves During Sinus Rhythm: Depiction of the Arrhythmogenic Substrate? *Circ. Arrhythmia Electrophysiol.* **2017**, *10*, e005145. [CrossRef] [PubMed]
6. Jadidi, A.; Nothstein, M.; Chen, J.; Lehrmann, H.; Dössel, O.; Allgeier, J.; Trenk, D.; Neumann, F.J.; Loewe, A.; Müller-Edenborn, B.; et al. Specific Electrogram Characteristics Identify the Extra-Pulmonary Vein Arrhythmogenic Sources of Persistent Atrial Fibrillation—Characterization of the Arrhythmogenic Electrogram Patterns During Atrial Fibrillation and Sinus Rhythm. *Sci. Rep.* **2020**, *10*, 9147. [CrossRef] [PubMed]
7. Kharbanda, R.K.; Knops, P.; van der Does, L.; Kik, C.; Taverne, Y.; Roos-Serote, M.C.; Heida, A.; Oei, F.; Bogers, A.; de Groot, N. Simultaneous Endo-Epicardial Mapping of the Human Right Atrium: Unraveling Atrial Excitation. *J. Am. Heart Assoc.* **2020**, *9*, e017069. [CrossRef]
8. Shah, D.; Haïssaguerre, M.; Jaïs, P. Catheter Ablation of Pulmonary Vein Foci for Atrial Fibrillation. *Thorac. Cardiovasc. Surg.* **1999**, *47*, 352–356. [CrossRef] [PubMed]
9. Oral, H.; Knight, B.P.; Tada, H. Pulmonary vein isolation for paroxysmal and persistent atrial fibrillation. *Circulation* **2002**, *11*, 83. [CrossRef]
10. Della Rocca, D.G.; Tarantino, N.; Trivedi, C.; Mohanty, S.; Anannab, A.; Salwan, A.S.; Gianni, C.; Bassiouny, M.; Al-Ahmad, A.; Romero, J.; et al. Non-pulmonary vein triggers in nonparoxysmal atrial fibrillation: Implications of pathophysiology for catheter ablation. *J. Cardiovasc. Electrophysiol.* **2020**, *31*, 2154–2167. [CrossRef]

11. Lin, W.S.; Tai, C.T.; Hsieh, M.H. Catheter ablation of paroxysmal atrial fibrillation initiated by non-pulmonary vein ectopy. *Circulation* **2003**, *12*, 53. [CrossRef]

12. Sánchez-Quintana, D.; López-Mínguez, J.R.; Pizarro, G.; Murillo, M.; Cabrera, J.A. Triggers and anatomical substrates in the genesis and perpetuation of atrial fibrillation. *Curr. Cardiol. Rev.* **2012**, *8*, 310–326. [CrossRef]

13. Nademanee, K.; McKenzie, J.; Kosar, E.; Schwab, M.; Sunsaneewitayakul, B.; Vasavakul, T.; Khunnawat, C.; Ngarmukos, T. A new approach for catheter ablation of atrial fibrillation: Mapping of the electrophysiologic substrate. *J. Am. Coll. Cardiol.* **2004**, *43*, 2044–2053. [CrossRef]

14. Vraka, A.; Hornero, F.; Bertomeu-González, V.; Osca, J.; Alcaraz, R.; Rieta, J.J. Short-Time Estimation of Fractionation in Atrial Fibrillation with Coarse-Grained Correlation Dimension for Mapping the Atrial Substrate. *Entropy* **2020**, *22*, 232. [CrossRef]

15. Chang, S.L.; Tai, C.T.; Lin, Y.J.; Wongcharoen, W.; Lo, L.W.; Tuan, T.C.; Udyavar, A.R.; Chang, S.H.; Tsao, H.M.; Hsieh, M.H.; et al. Biatrial substrate properties in patients with atrial fibrillation. *J. Cardiovasc. Electrophysiol.* **2007**, *18*, 1134–1139. [CrossRef] [PubMed]

16. Garg, L.; Pothineni, N.V.K.; Daw, J.M.; Hyman, M.C.; Arkles, J.; Tschabrunn, C.M.; Santangeli, P.; Marchlinski, F.E. Impact of Left Atrial Bipolar Electrogram Voltage on First Pass Pulmonary Vein Isolation During Radiofrequency Catheter Ablation. *Front. Physiol.* **2020**, *11*, 594654. [CrossRef] [PubMed]

17. Verma, A.; Jiang, C.y.; Betts, T.R.; Chen, J.; Deisenhofer, I.; Mantovan, R.; Macle, L.; Morillo, C.A.; Haverkamp, W.; Weerasooriya, R.; et al. Approaches to catheter ablation for persistent atrial fibrillation. *N. Engl. J. Med.* **2015**, *372*, 1812–1822. [CrossRef] [PubMed]

18. Cheng, W.H.; Lo, L.W.; Lin, Y.J.; Chang, S.L.; Hu, Y.F.; Hung, Y.; Chung, F.P.; Liao, J.N.; Tuan, T.C.; Chao, T.F.; et al. Ten-year ablation outcomes of patients with paroxysmal atrial fibrillation undergoing pulmonary vein isolation. *Heart Rhythm* **2019**, *16*, 1327–1333. [CrossRef]

19. Inamura, Y.; Nitta, J.; Inaba, O.; Sato, A.; Takamiya, T.; Murata, K.; Ikenouchi, T.; Kono, T.; Matsumura, Y.; Takahashi, Y.; et al. Presence of non-pulmonary vein foci in patients with atrial fibrillation undergoing standard ablation of pulmonary vein isolation: Clinical characteristics and long-term ablation outcome. *Int. J. Cardiol. Heart Vasc.* **2021**, *32*, 100717. [CrossRef] [PubMed]

20. Simpson, R.J.; Foster, J.R.; Gettes, L.S. Atrial excitability and conduction in patients with interatrial conduction defects. *Am. J. Cardiol.* **1982**, *50*, 1331–1337. [CrossRef]

21. Blanche, C.; Tran, N.; Rigamonti, F.; Burri, H.; Zimmermann, M. Value of P-wave signal averaging to predict atrial fibrillation recurrences after pulmonary vein isolation. *EP Eur.* **2013**, *15*, 198–204. [CrossRef]

22. Chen, Q.; Mohanty, S.; Trivedi, C.; Gianni, C.; Della Rocca, D.G.; Canpolat, U.; Burkhardt, J.D.; Sanchez, J.E.; Hranitzky, P.; Gallinghouse, G.J.; et al. Association between prolonged P wave duration and left atrial scarring in patients with paroxysmal atrial fibrillation. *J. Cardiovasc. Electrophysiol.* **2019**, *30*, 1811–1818. [CrossRef] [PubMed]

23. Pranata, R.; Yonas, E.; Vania, R. Prolonged P-wave duration in sinus rhythm pre-ablation is associated with atrial fibrillation recurrence after pulmonary vein isolation—A systematic review and meta-analysis. *Ann. Noninvasive Electrocardiol.* **2019**, *24*, e12653. [CrossRef] [PubMed]

24. Auricchio, A.; Özkartal, T.; Salghetti, F.; Neumann, L.; Pezzuto, S.; Gharaviri, A.; Demarchi, A.; Caputo, M.L.; Regoli, F.; De Asmundis, C.; et al. Short P-Wave Duration is a Marker of Higher Rate of Atrial Fibrillation Recurrences after Pulmonary Vein Isolation: New Insights into the Pathophysiological Mechanisms Through Computer Simulations. *J. Am. Heart Assoc.* **2021**, *10*, e018572. [CrossRef] [PubMed]

25. Miao, Y.; Xu, M.; Yang, L.; Zhang, C.; Liu, H.; Shao, X. Investigating the association between P wave duration and atrial fibrillation recurrence after radiofrequency ablation in early persistent atrial fibrillation patients. *Int. J. Cardiol.* **2021**, *351*, 48–54. [CrossRef] [PubMed]

26. Van Beeumen, K.; Houben, R.; Tavernier, R.; Ketels, S.; Duytschaever, M. Changes in P-wave area and P-wave duration after circumferential pulmonary vein isolation. *EP Eur.* **2010**, *12*, 798–804. [CrossRef] [PubMed]

27. Maan, A.; Mansour, M.; Ruskin, J.N.; Heist, E.K. Impact of catheter ablation on P-wave parameters on 12-lead electrocardiogram in patients with atrial fibrillation. *J. Electrocardiol.* **2014**, *47*, 725–733. [CrossRef]

28. Hu, X.; Jiang, J.; Ma, Y.; Tang, A. Novel P Wave Indices to Predict Atrial Fibrillation Recurrence After Radiofrequency Ablation for Paroxysmal Atrial Fibrillation. *Med. Sci. Monit.* **2016**, *22*, 2616–2623. [CrossRef] [PubMed]

29. Vraka, A.; Bertomeu-González, V.; Hornero, F.; Quesada, A.; Alcaraz, R.; Rieta, J.J. Splitting the P-Wave: Improved Evaluation of Left Atrial Substrate Modification after Pulmonary Vein Isolation of Paroxysmal Atrial Fibrillation. *Sensors* **2022**, *22*, 290. [CrossRef] [PubMed]

30. Alcaraz, R.; Martínez, A.; Rieta, J.J. The P Wave Time-Frequency Variability Reflects Atrial Conduction Defects before Paroxysmal Atrial Fibrillation. *Ann. Noninvasive Electrocardiol.* **2015**, *20*, 433–445. [CrossRef]

31. Murase, Y.; Imai, H.; Ogawa, Y.; Kano, N.; Mamiya, K.; Ikeda, T.; Okabe, K.; Arai, K.; Yamazoe, S.; Torii, J.; et al. Usefulness of P-wave duration in patients with sick sinus syndrome as a predictor of atrial fibrillation. *J. Arrhythmia* **2021**, *37*, 1220–1226. [CrossRef] [PubMed]

32. Wu, J.T.; Dong, J.Z.; Sang, C.H.; Tang, R.B.; Ma, C.S. Prolonged PR interval and risk of recurrence of atrial fibrillation after catheter ablation. *Int. Heart J.* **2014**, *55*, 126–130. [CrossRef]

33. Salah, A.; Zhou, S.; Liu, Q.; Yan, H. P wave indices to predict atrial fibrillation recurrences post pulmonary vein isolation. *Arq. Bras. Cardiol.* **2013**, *101*, 519–527. [CrossRef] [PubMed]

34. Alcaraz, R.; Martínez, A.; Rieta, J.J. Role of the P-wave high frequency energy and duration as noninvasive cardiovascular predictors of paroxysmal atrial fibrillation. *Comput. Methods Programs Biomed.* **2015**, *119*, 110–119. [CrossRef]

35. Task Force of the European Society of Cardiology the North American Society of Pacing Electrophysiologys. Heart rate variability. Standards of measurement, physiological interpretation, and clinical use. *Eur. Heart J.* **1996**, *17*, 354–381. [CrossRef]

36. Perkiömäki, J.; Ukkola, O.; Kiviniemi, A.; Tulppo, M.; Ylitalo, A.; Kesäniemi, Y.A.; Huikuri, H. Heart rate variability findings as a predictor of atrial fibrillation in middle-aged population. *J. Cardiovasc. Electrophysiol.* **2014**, *25*, 719–724. [CrossRef] [PubMed]

37. Habibi, M.; Chahal, H.; Greenland, P.; Guallar, E.; Lima, J.A.C.; Soliman, E.Z.; Alonso, A.; Heckbert, S.R.; Nazarian, S. Resting Heart Rate, Short-Term Heart Rate Variability and Incident Atrial Fibrillation (from the Multi-Ethnic Study of Atherosclerosis (MESA)). *Am. J. Cardiol.* **2019**, *124*, 1684–1689. [CrossRef] [PubMed]

38. Hsieh, M.H.; Chiou, C.W.; Wen, Z.C.; Wu, C.H.; Tai, C.T.; Tsai, C.F.; Ding, Y.A.; Chang, M.S.; Chen, S.A. Alterations of heart rate variability after radiofrequency catheter ablation of focal atrial fibrillation originating from pulmonary veins. *Circulation* **1999**, *100*, 2237–2243. [CrossRef]

39. Chen, P.S.; Chen, L.S.; Fishbein, M.C.; Lin, S.F.; Nattel, S. Role of the autonomic nervous system in atrial fibrillation: Pathophysiology and therapy. *Circ. Res.* **2014**, *114*, 1500–1515. [CrossRef] [PubMed]

40. Zhu, Z.; Wang, W.; Cheng, Y.; Wang, X.; Sun, J. The predictive value of heart rate variability indices tested in early period after radiofrequency catheter ablation for the recurrence of atrial fibrillation. *J. Cardiovasc. Electrophysiol.* **2020**, *31*, 1350–1355. [CrossRef]

41. Maille, B.; Das, M.; Hussein, A.; Shaw, M.; Chaturvedi, V.; Williams, E.; Morgan, M.; Ronayne, C.; Snowdon, R.L.; Gupta, D. Reverse electrical and structural remodeling of the left atrium occurs early after pulmonary vein isolation for persistent atrial fibrillation. *J. Interv. Card. Electrophysiol.* **2020**, *58*, 9–19. [CrossRef] [PubMed]

42. Daoud, E.G.; Niebauer, M.; Bakr, O.; Jentzer, J.; Man, K.C.; Williamson, B.D.; Hummel, J.D.; Strickberger, S.A.; Morady, F. Placement of electrode catheters into the coronary sinus during electrophysiology procedures using a femoral vein approach. *Am. J. Cardiol.* **1994**, *74*, 194–195. [CrossRef]

43. Antz, M.; Otomo, K.; Arruda, M.; Scherlag, B.J.; Pitha, J.; Tondo, C.; Lazzara, R.; Jackman, W.M. Electrical conduction between the right atrium and the left atrium via the musculature of the coronary sinus. *Circulation* **1998**, *98*, 1790–1795. [CrossRef] [PubMed]

44. Santangeli, P.; Marchlinski, F.E. Techniques for the provocation, localization, and ablation of non–pulmonary vein triggers for atrial fibrillation. *Heart Rhythm* **2017**, *14*, 1087–1096. [CrossRef] [PubMed]

45. Ahmed, N.; Perveen, S.; Mehmood, A.; Rani, G.F.; Molon, G. Coronary Sinus Ablation Is a Key Player Substrate in Recurrence of Persistent Atrial Fibrillation. *Cardiology* **2019**, *143*, 107–113. [CrossRef]

46. Razeghian-Jahromi, I.; Natale, A.; Nikoo, M.H. Coronary sinus diverticulum: Importance, function, and treatment. *Pacing Clin. Electrophysiol. PACE* **2020**, *43*, 1582–1587. [CrossRef]

47. Boles, U.; Gul, E.E.; Enriquez, A.; Starr, N.; Haseeb, S.; Abdollah, H.; Simpson, C.; Baranchuk, A.; Redfearn, D.; Michael, K.; et al. Coronary sinus electrograms may predict new-onset atrial fibrillation after typical atrial flutter radiofrequency ablation. *J. Atr. Fibrillation* **2018**, *11*, 1809. [CrossRef] [PubMed]

48. Morita, H.; Zipes, D.P.; Morita, S.T.; Wu, J. The role of coronary sinus musculature in the induction of atrial fibrillation. *Heart Rhythm* **2012**, *9*, 581–589. [CrossRef] [PubMed]

49. Tahara, M.; Kato, R.; Ikeda, Y.; Goto, K.; Asano, S.; Mori, H.; Iwanaga, S.; Muramatsu, T.; Matsumoto, K. Differential Atrial Pacing to Detect Reconnection Gaps After Pulmonary Vein Isolation in Atrial Fibrillation. *Int. Heart J.* **2020**, *61*, 503–509. [CrossRef]

50. Mahmud, E.; Raisinghani, A.; Keramati, S.; Auger, W.; Blanchard, D.G.; DeMaria, A.N. Dilation of the coronary sinus on echocardiogram: Prevalence and significance in patients with chronic pulmonary hypertension. *JASE* **2001**, *14*, 44–49. [CrossRef] [PubMed]

51. Langenberg, C.J.M.; Pietersen, H.G.; Geskes, G.; Wagenmakers, A.J.M.; Soeters, P.B.; Durieux, M. Coronary Sinus Catheter Placement. *Clin. Investig. Cardiol.* **2003**, *124*, 1259–1265. [CrossRef]

52. Saremi, F.; Thonar, B.; Sarlaty, T.; Shmayevich, I.; Malik, S.; Smith, C.W.; Krishnan, S.; Sánchez-Quintana, D.; Narula, N. Posterior interatrial muscular connection between the coronary sinus and left atrium: Anatomic and functional study of the coronary sinus with multidetector CT. *Radiology* **2011**, *260*, 671–679. [CrossRef] [PubMed]

53. Vogt, J.; Heintze, J.; Hansky, B.; Güldner, H.; Buschler, H.; Horstkotte, D. Implantation: Tips and tricks–the cardiologist's view. *Eur. Heart J. Suppl.* **2004**, *6*, D47–D52. [CrossRef]

54. Pai, R.G.; Varadarajan, P.; Tanimoto, M. Effect of atrial fibrillation on the dynamics of mitral annular area. *J. Heart Valve Dis.* **2003**, *12*, 31–37. [PubMed]

55. El-Maasarany, S.; Ferrett, C.G.; Firth, A.; Sheppard, M.; Henein, M.Y. The coronary sinus conduit function: Anatomical study (relationship to adjacent structures). *EP Eur.* **2005**, *7*, 475–481. [CrossRef]

56. Meek, S.; Morris, F. ABC of clinical electrocardiography: Introduction. II—basic terminology. *BMJ* **2002**, *324*, 470–473. [CrossRef]

57. García, M.; Martínez-Iniesta, M.; Ródenas, J.; Rieta, J.J.; Alcaraz, R. A novel wavelet-based filtering strategy to remove powerline interference from electrocardiograms with atrial fibrillation. *Physiol. Meas.* **2018**, *39*, 115006. [CrossRef] [PubMed]

58. Sörnmo, L.; Laguna, P. Electrocardiogram (ECG) Signal Processing. In *Wiley Encyclopedia of Biomedical Engineering*; John Wiley and Sons: Hoboken, NJ, USA, 2006; Volume 2, pp. 1298–1313. [CrossRef]

59. Martínez-Iniesta, M.; Ródenas, J.; Rieta, J.J.; Alcaraz, R. The stationary wavelet transform as an efficient reductor of powerline interference for atrial bipolar electrograms in cardiac electrophysiology. *Physiol. Meas.* **2019**, *40*, 075003. [CrossRef] [PubMed]

60. Alcaraz, R.; Rieta, J.J. Adaptive singular value cancelation of ventricular activity in single-lead atrial fibrillation electrocardiograms. *Physiol. Meas.* **2008**, *29*, 1351–1369. [CrossRef] [PubMed]
61. Martínez, A.; Alcaraz, R.; Rieta, J.J. Detection and removal of ventricular ectopic beats in atrial fibrillation recordings via principal component analysis. In Proceedings of the Annual International Conference of the IEEE Engineering in Medicine and Biology Society, Boston, MA, USA, 30 August–3 September 2011; pp. 4693–4696. [CrossRef]
62. Choi, A.; Shin, H. Quantitative Analysis of the Effect of an Ectopic Beat on the Heart Rate Variability in the Resting Condition. *Front. Physiol.* **2018**, *9*, 922. [CrossRef]
63. Martinez, A.; Alcaraz, R.; Rieta, J.J. A new method for automatic delineation of ECG fiducial points based on the Phasor Transform. In Proceedings of the Annual International Conference of the IEEE Engineering in Medicine and Biology Society, Buenos Aires, Argentina, 31 August–4 September 2010; pp. 4586–4589. [CrossRef]
64. González, F.; Alcaraz, R.; Rieta, J.J. Electrocardiographic P-wave delineation based on adaptive slope Gaussian detection. In Proceedings of the 2017 Computing in Cardiology (CinC), Rennes, France, 24–27 September 2017; pp. 1–4. [CrossRef]
65. Osorio, D.; Alcaraz, R.; Rieta, J.J. A fractionation-based local activation wave detector for atrial electrograms of atrial fibrillation. In Proceedings of the 2017 Computing in Cardiology (CinC), Rennes, France, 24–27 September 2017; pp. 1–4. [CrossRef]
66. Vraka, A.; Bertomeu-González, V.; Osca, J.; Ravelli, F.; Alcaraz, R.; Rieta, J.J. Study on How Catheter Ablation Affects Atrial Structures in Patients with Paroxysmal Atrial Fibrillation: The Case of the Coronary Sinus. In Proceedings of the 2020 International Conference on e-Health and Bioengineering (EHB), Iasi, Romania, 29–30 October 2020; pp. 1–4. [CrossRef]
67. Alcaraz, R.; Hornero, F.; Martínez, A.; Rieta, J.J. Short-time regularity assessment of fibrillatory waves from the surface ECG in atrial fibrillation. *Physiol. Meas.* **2012**, *33*, 969–984. [CrossRef] [PubMed]
68. Zawadzki, J.; Zawadzki, G.; Radziejewska, J.; Wolff, P.S.; Sławuta, A.; Gajek, J. The P wave dispersion—One pixel, one millisecond. *Rev. Cardiovasc. Med.* **2021**, *22*, 1633. [CrossRef] [PubMed]
69. Toman, O.; Hnatkova, K.; Smetana, P.; Huster, K.M.; Šišáková, M.; Barthel, P.; Novotný, T.; Schmidt, G.; Malik, M. Physiologic heart rate dependency of the PQ interval and its sex differences. *Sci. Rep.* **2020**, *10*, 2551. [CrossRef] [PubMed]
70. Mandelbrot, B.B. Contributions to Probability and Statistics: Essays in Honor of Harold Hotelling. *Siam Rev.* **1961**, *3*, 80.
71. Shapiro, S.S.; Wilk, M.B., An Analysis of Variance Test for Normality (Complete Samples). *Biometrika* **1965**, *52*, 591–611. [CrossRef]
72. Kruskal, W.H.; Wallis, W.A. Use of ranks in one-criterion variance analysis. *J. Am. Stat. Assoc.* **1952**, *47*, 583–621. [CrossRef]
73. Mann, H.B.; Whitney, D.R. On a Test of Whether one of Two Random Variables is Stochastically Larger than the Other. *Ann. Math. Stat.* **1947**, *18*, 50–60. [CrossRef]
74. Oral, H.; Ozaydin, M.; Chugh, A.; Scharf, C.; Tada, H.; Hall, B.; Cheung, P.; Pelosi, F.; Knight, B.P.; Morady, F. Role of the coronary sinus in maintenance of atrial fibrillation. *J. Cardiovasc. Electrophysiol.* **2003**, *14*, 1329–1336. [CrossRef]
75. Di Marco, L.Y.; Raine, D.; Bourke, J.P.; Langley, P. Characteristics of atrial fibrillation cycle length predict restoration of sinus rhythm by catheter ablation. *Heart Rhythm* **2013**, *10*, 1303–1310. [CrossRef]
76. Misek, J.; Belyaev, I.; Jakusova, V.; Tonhajzerova, I.; Barabas, J.; Jakus, J. Heart rate variability affected by radiofrequency electromagnetic field in adolescent students. *Bioelectromagnetics* **2018**, *39*, 277–288. [CrossRef] [PubMed]
77. Misek, J.; Veternik, M.; Tonhajzerova, I.; Jakusova, V.; Janousek, L.; Jakus, J. Radiofrequency Electromagnetic Field Affects Heart Rate Variability in Rabbits. *Physiol. Res.* **2020**, *69*, 633–643. [CrossRef] [PubMed]
78. Ortigosa, N.; Ayala, G.; Cano, Ó. Variation of P-wave indices in paroxysmal atrial fibrillation patients before and after catheter ablation. *Biomed. Signal Process. Control* **2021**, *66*, 102500. [CrossRef]
79. McCann, A.; Vesin, J.M.; Pruvot, E.; Roten, L.; Sticherling, C.; Luca, A. ECG-Based Indices to Characterize Persistent Atrial Fibrillation Before and During Stepwise Catheter Ablation. *Front. Physiol.* **2021**, *12*, 654053. [CrossRef] [PubMed]
80. Knecht, S.; Pradella, M.; Reichlin, T.; Mühl, A.; Bossard, M.; Stieltjes, B.; Conen, D.; Bremerich, J.; Osswald, S.; Kühne, M.; et al. Left atrial anatomy, atrial fibrillation burden, and P-wave duration—Relationships and predictors for single-procedure success after pulmonary vein isolation. *EP Eur.* **2018**, *20*, 271–278. [CrossRef]

Review

Atrial Fibrillation Specific Exercise Rehabilitation: Are We There Yet?

Benjamin J. R. Buckley [1,2,*], Signe S. Risom [3,4,5], Maxime Boidin [6], Gregory Y. H. Lip [1,2,6] and Dick H. J. Thijssen [6,7]

[1] Liverpool Centre for Cardiovascular Science, University of Liverpool, William Henry Duncan Building, Liverpool L7 8TX, UK; gregory.lip@liverpool.ac.uk

[2] Cardiovascular and Metabolic Medicine, Institute of Life Course and Medical Sciences, University of Liverpool, Liverpool L7 8TX, UK

[3] Department of Cardiology, Herlev and Gentofte University Hospital, 2730 Herlev, Denmark; signe.stelling.risom@regionh.dk

[4] Institute of Nursing and Nutrition, University College Copenhagen, 1799 Copenhagen, Denmark

[5] Department of Clinical Medicine, Faculty of Health and Medical Sciences, University of Copenhagen, 1799 Copenhagen, Denmark

[6] Liverpool Centre for Cardiovascular Science, Liverpool John Moores University, Liverpool L3 3AF, UK; m.boidin@ljmu.ac.uk (M.B.); dick.thijssen@radboudumc.nl (D.H.J.T.)

[7] Department of Physiology, Research Institute for Health Science, Radboud University Medical Centerum, P.O. Box 9101, 6500 HB Nijmegen, The Netherlands

* Correspondence: benjamin.buckley@liverpool.ac.uk; Tel.: +44-(0)151-794-2000

Abstract: Regular physical activity and exercise training are integral for the secondary prevention of cardiovascular disease. Despite recent advances in more holistic care pathways for people with atrial fibrillation (AF), exercise rehabilitation is not provided as part of routine care. The most recent European Society of Cardiology report for AF management states that patients should be encouraged to undertake moderate-intensity exercise and remain physically active to prevent AF incidence or recurrence. The aim of this review was to collate data from primary trials identified in three systematic reviews and recent real-world cohort studies to propose an AF-specific exercise rehabilitation guideline. Collating data from 21 studies, we propose that 360–720 metabolic equivalent (MET)-minutes/week, corresponding to ~60–120 min of exercise per week at moderate-to-vigorous intensity, could be an evidence-based recommendation for patients with AF to improve AF-specific outcomes, quality of life, and possibly prevent long-term major adverse cardiovascular events. Furthermore, non-traditional, low-moderate intensity exercise, such as Yoga, seems to have promising benefits on patient quality of life and possibly physical capacity and should, therefore, be considered in a personalised rehabilitation programme. Finally, we discuss the interesting concepts of short-term exercise-induced cardioprotection and 'none-response' to exercise training with reference to AF rehabilitation.

Keywords: rehabilitation medicine; physical activity; exercise; atrial fibrillation; cardiovascular disease; preventive cardiology; vascular health; atrial health; secondary prevention

Citation: Buckley, B.J.R.; Risom, S.S.; Boidin, M.; Lip, G.Y.H.; Thijssen, D.H.J. Atrial Fibrillation Specific Exercise Rehabilitation: Are We There Yet?. *J. Pers. Med.* **2022**, *12*, 610. https://doi.org/10.3390/jpm12040610

Academic Editor: Chiara Bellia

Received: 17 February 2022
Accepted: 31 March 2022
Published: 10 April 2022

1. Why Do We Need AF-Specific, Exercise-Based Rehabilitation?

Regular physical activity and exercise training are integral for the secondary prevention of cardiovascular disease (CVD), as demonstrated in both interventional and real-world studies [1–4]. However, 42% of general western populations do not meet the recommended physical activity guidelines (i.e., 150 min of moderate or 75 min of vigorous intensity physical activity/week) [5]. Further, those with CVD are typically less active than the general population yet stand to benefit the most from exercise training [6]. For example, Jeong et al. [6] demonstrated that, in 131,558 individuals with CVD, every 500 metabolic equivalent (MET)-minute/week increase in physical activity resulted in a 14% risk reduction in mortality, whereas in 310,240 participants without CVD, risk reduction was only 7%. Interestingly, while individuals without CVD benefited the most between 1 and 500 METs-min/week of physical activity, the benefit in those with CVD continued above 500−1000 METs-min/week. Thus, exercise interventions, such as cardiac rehabilitation, are an essential component of secondary and tertiary cardiovascular disease management. It is also apparent that for rehabilitation programmes, one size does not fit all [7], and optimised interventions should be developed to enhance outcomes for different population groups.

There is increasing interest for a more personalised approach to exercise-based cardiovascular rehabilitation, especially since cardiac (chronotropic incompetence, diastolic dysfunction, and systolic dysfunction), non-cardiac (vascular function/structure, skeletal myopathy, and autonomic control), comorbidities (ageing, metabolic diseases, and cardiovascular diseases), and external (exercise adherence/dose/intensity) factors influence individual responses [8]. In particular, these factors have been shown to moderate baseline and even the relative change in cardiorespiratory fitness ($\dot{V}O_2$-peak) following exercise training [8].

From the cardiac and comorbidity perspective, there are specific exercise guidelines for a number of cardiovascular conditions, including ischaemic heart disease, heart failure, hypertension, and peripheral artery disease [9,10]. These condition-specific guidelines help to personalise exercise prescription at a group level and represent an initial step towards the optimisation of individual patient benefit.

The most recent European Society of Cardiology (ESC) guidelines for the diagnosis and management of atrial fibrillation (AF) repeatedly highlight the importance of cardiopulmonary exercise testing (CPET) and cardiorespiratory fitness, as well as stating that *'patients should be encouraged to undertake moderate-intensity exercise and remain physically active to prevent AF incidence or recurrence'* [11]. Furthermore, the latest ESC Guidelines on Sports Cardiology and Exercise in Patients with Cardiovascular Disease emphasise the primary and secondary preventive effects of physical activity and present specific guidelines for healthy individuals and those with cardiovascular disease risk factors [9]. Specific secondary preventive exercise guidelines are, however, not provided for people with AF.

Exercise rehabilitation presents an important and potentially impactful initiative for people with AF and the health services they use. However, more specific exercise-rehabilitation guidelines are needed for people with AF. We, therefore, conducted a narrative review of the literature, supported by recent systematic reviews and primary studies, to produce more nuanced exercise rehabilitation thresholds for people with AF. By working towards such an objective, we can better guide both healthcare professionals and patients by recommending AF-specific exercise-based rehabilitation and/or guide future research to help contribute to this research gap.

2. Methods

Original trials were included from a Cochrane systematic review of exercise-based cardiac rehabilitation [12] and three more recent systematic reviews investigating exercise-based cardiac rehabilitation [13,14] and different types of exercise interventions [15] for adults with AF. In addition, we included recent and relevant prospective and retrospective studies that investigated the impact of exercise or physical activity on health outcomes for people with AF.

It is well known that there is a seemingly counterintuitive increased risk of AF for some who engage in excessive amounts of vigorous-intensity exercise (i.e., 'athletic AF') [16]. However, athletic AF is only observed following exceptionally high levels of vigorous endurance training, such as >5000 MET-min/week or 5 to 10-fold the existing physical activity guidelines [16]. Given that athletic AF represents a relatively small percentage of the AF population, the present review will focus on the impact of physical activity and exercise training on the secondary prevention of AF and associated outcomes in non-athletic populations only.

Data from relevant and eligible primary studies were extracted by one reviewer to define the AF population, exercise intervention or physical activity behaviour details (i.e., frequency, intensity, time, and type (FITT) principles), follow-up time points, and subsequent health outcomes, stratified by study design (Table 1). All extracted data were quality checked for accuracy by a second reviewer. Given the heterogeneity in investigated outcome measures across the literature, we decided to focus on a priori serious adverse events (SAE; mortality, hospitalization, and stroke), physical capacity (including the 6 min walk test and $\dot{V}O_2$-peak), AF specific outcomes (including recurrence and/or time in AF), and quality of life (QoL). These data were then used to determine if more nuanced exercise recommendations could be proposed and highlight where further research is needed. Aligned with the extracted data regarding AF-specific exercise rehabilitation, we then discuss the hypothesised mechanisms of long-term (Figure 1) and possible acute AF protection following exercise training and physical activity. Finally, we conclude with key take-home messages and recommendations for future research in this area.

Table 1. Intervention components/physical activity profile and outcomes of interest from included studies.

First Author, Year	AF Population *n* = Sample; Age (Years); Sex (%); AF Subtype	Exercise Intervention *Frequency; Intensity; Time; Type*	Follow-Up and Impact on Health Outcomes (SAE, Physical Capacity, AF Specific Outcomes, and QoL)
	Randomised Controlled Trials		
Luo, 2017 [17]	*n* = 382 63 years 16% female AF and heart failure (excluded if sustained fast AF)	*Frequency*: 3 sessions/week for 36 sessions, followed by transition to a home-based exercise program for 2 years. *Intensity*: NR *Time*: 90 min/week for 3 months, followed by 120 min/week thereafter *Type*: Aerobic exercise (walking, treadmill, or cycle ergometer)	Follow-up: median of 2.6 years SAE: AF was associated with a 24% per year higher rate of mortality/hospitalisation in the control group compared to intervention group (HR: 1.53; 95% CI: 1.34 to 1.74; *p* < 0.001) in unadjusted analysis; this did not remain significant after adjustment (HR: 1.15;95% CI: 0.98 to 1.35; *p* > 0.09). No interaction between AF and exercise training (all *p* > 0.10). No difference in AF event rates between groups (all *p* > 0.10).
Malmo, 2016 [18]	*n* = 51 Intervention: 56 years 77% female Control: 62 years 88% male Paroxysmal or persistent AF	*Frequency*: 3 sessions/week for 12 weeks *Intensity*: Vigorous (85–95% of maximal heart rate and Borg 6–20) *Time*: 4 min intervals *Type*: Walking or running on treadmill	Follow-up: 20 weeks Mean time in AF: 10.4% (95% CI, 4.6–17.8) to 14.6% (95% CI, 6.4–24.9) in the control group vs. 8.1% (95% CI, 4.1–12.8) to 4.8% (95% CI, 2.0–7.6) in the exercise group (*p* = 0.001). $\dot{V}O_2$-peak change to follow-Up: Control group: −0.3 ± 4.3 vs. Exercise group: 3.2 ± 2.5, *p* < 0.001. Quality of life change to Follow-Up, SF-36: Physical component score: Control group: −0.3 ± 5.4 vs. exercise group: 2.2 ± 4.4 Mental component score: Control group: 1.4 ± 7.2 vs. Exercise group: 3.6 ± 6.5, *p* > 0.05. No serious adverse events, 2 patients in EG had bursitis episodes, 2 patients in CG had a stroke and ventricular tachycardia.
Osbak, 2011	*n* = 49 Intervention: 70 years Control: 71 years 75% male Permanent AF	*Frequency*: 3 sessions/week for 12 weeks *Intensity*: Vigorous (70% of maximal exercise capacity or Borg scale 14–16/20) *Time*: 1 h *Type*: Group-based aerobic training including ergometer cycling, walking on stairs, running, fitness training on physioballs, and interval training.	Follow-up: 12 weeks Physical capacity: 6MWT: Significantly increased within the intervention group (from 504.4 (85.1) m to 569.9 (92.6) m (*p* < 0.001) and between the groups (EG = 569.9 (92.6) m and CG = 454.1 (95.7), (*p* = 0.001). QoL: MLHF-Q score: intervention group: 21.1 ± 18.0 vs. control group: 15.4 ± 17.5, *p* = 0.03. SF-36 subscales: physical functioning (*p* = 0.02), general health perceptions (*p* = 0.001), and vitality (*p* = 0.02) in favour of the intervention group. No adverse events
Pippa, 2007	*n* = 43 Intervention: 68 years 64% male Control: 68 years 76% male Permanent AF	*Frequency*: 2 sessions/week for 16 weeks *Intensity*: NR (predict light) *Time*: 90 min *Type*: Qigong training (consists of slow and graceful movements with a focus on breathing)	Follow-up: 16 weeks Physical capacity: 6MWT: Intervention group: 531(121) meters at the end of intervention, 474 (109) meters at 16 weeks after intervention vs. control group: 380 (97) meters at the end of intervention, 350 (110) meters after 16 weeks. *p* < 0.001. 1 retinal embolism in the intervention and 1 case of deep vein thrombosis during the follow-up in the control group.
Rienstra, 2018	*n* = 245 Intervention: 64 years 79% male Control: 65 years 79% male Persistent AF	*Frequency*: 2–3 sessions/week for 9–11 weeks *Intensity*: NR *Time*: 20–30 min exercise, *Type*: Cardiac rehabilitation	Follow-up: 1 year AF specific: At 1 year, sinus rhythm was present in 89 (75%) patients in the intervention vs. 79 (63%) in the conventional group (odds ratio 1.765, lower limit of 95% confidence interval 1.021, *p* = 0.042).

Table 1. *Cont.*

First Author, Year	AF Population *n* = Sample; Age (Years); Sex (%); AF Subtype	Exercise Intervention *Frequency; Intensity; Time; Type*	Follow-Up and Impact on Health Outcomes (SAE, Physical Capacity, AF Specific Outcomes, and QoL)
Risom, 2016 Long-term follow up: Risom, 2020	*n* = 210 Intervention: 60 years Male 70% Control: 59 years 73% male Paroxysmal or persistent AF	*Frequency*: 3 sessions/week for 12 weeks *Intensity*: Intensity was progressively increased *Type*: Comprehensive cardiac rehabilitation. Cardiovascular training and strength exercises	Follow-up: 6 and 12-months SAE: Mortality: one death in each group at 6- and 24-months follow-up ($p > 0.99$). All hospital admissions: Intervention group: $n = 71$ (68%) vs. control group: $n = 60$ (57%). Physical capacity: $\dot{V}O_2$-peak at four months: (Intervention group: 24.3 mL kg^{-1} min^{-1} vs control group: 20.7 mL kg^{-1} min^{-1}, $p = 0.003$). VO_2 peak at twelve months: (Intervention group: 25.8 mL kg^{-1} min^{-1} vs control group: 22.4 mL kg^{-1} min^{-1}, $p = 0.002$). OoL: SF-36: General health Perception (Intervention group: 67.16 points vs. control group: 66.9 points. $p = 0.02$, of Interaction between Intervention and time). No significant difference between groups was found at six or 24 at the other domains. AFEQT: The results are in favour of the intervention group: Global score: (Intervention group: 81.64 points vs. control group: 82.87 points $p = 0.04$, of Interaction between Intervention and time) and the treatment satisfaction $p = 0.03$, of Interaction between Intervention and time score at 24 months. Two serious adverse events (AF intervention- related and unrelated to intervention death in the EG and 1 unrelated to intervention death in the CG group. 16 non-serious adverse events EG and 7 in the CG.
Skielboe, 2017	*n* = 76 Low intensity: 64 years 58% male High intensity: 61 years 59% male Paroxysmal or persistent AF	*Frequency*: Two sessions/week for 12 weeks *Intensity*: exercise at either low or high intensity (50% (Borg scale 11–13/20) and 80% (Borg scale 16–18/20) of maximal perceived exertion, respectively) *Time*: 60 min *Type*: 20 min interval exercising on ergometer bike, 20 min varying circuit exercise on the floor.	Follow-up: 16 weeks. SAE: All hospital admissions: 19 patients in each group. Physical capacity: $\dot{V}O_2$-peak: No difference between groups: Mean diff. -0.76 mL O2/kg/min, 95% CI -3.22 ± 1.70. AF specific outcomes: Burden of AF measured by daily electrocardiography-reporting for 12 weeks. Results: No statistical difference between low and high intensity exercise for both unadjusted (IRR 0.983, 95% CI 0.39–2.46, $p = 0.971$) and adjusted analyses (IRR 0.742, 95% CI 0.29–1.91, $p = 0.538$). No serious adverse events in both groups. Three unserious adverse events reported in low intensity group and 5 in high intensity group, including symptoms of arrhythmia, hospital admission, AF ahead of an exercise session, and noncardiac complaints.
Wahlström, 2017	*n* = 80 Intervention: 64 years 48% male Control: 63 years 72%male Paroxysmal AF	*Frequency*: One session/week for 12 weeks Intensity: NR (predict light) *Time*: 30 min *Type*: Mediyoga (a therapeutic form of yoga evolved from Kundalini yoga). It is calm, meditative yoga based on deep breathing.	Follow-up: 14 weeks QoL: SF-36, Mental component scale improved in the intervention but not control: Intervention group: baseline 42.1 (17.6–53.5) to follow-up 50.6 (24.0–55.2) points vs. control group: baseline 53.0) 14.7–56.0) to follow-up 52.7 (24.5–57.1) points, $p = 0.016$. Physical component scale no difference within or between groups: Intervention group: 50.2 (27.6–59.1) points vs. control group: 49.0 (29.1–61.6) points, $p = 0.837$.
Joensen, 2019	*n* = 52 Intervention: 62 years 61% male Control: 60 years 71% male Paroxysmal or persistent AF	*Frequency*: 2 sessions/week for 12 weeks *Intensity*: Moderate-to-vigorous intensity ($\geq$ 70% of maximum exercise capacity or 14–16 on the Borg scale) *Time*: 60 min *Type*: Cardiac rehabilitation	Follow-up: 3, 6 and 12-months Physical capacity: Maximum exercise capacity improved in the intervention group from baseline (176 W (SD 48)) to 6 months (190 W (SD 55)). There was no change in the control group. 6MWT was improved in the EG form 613 m (96) to 644 m (84) with no statistically significant differences within or between the groups. AF-QoL-18 significantly improved in the intervention group from 48.4(22.8) to 68.0(15.2) compared with the control group (baseline 51.6 (SD 22.3), 6 months 59.2 (SD 27.3), $p = 0.031$. There was no statistical difference at 12-months. No statistical difference in AFEQT and EQ-VAS between intervention and control. 20 readmissions for cardiac reasons (mostly AF) in the intervention group and 18 in the control group. 13 direct current cardioversions in the intervention group and 12 in the control group. 7 radiofrequency ablations in the intervention group, and 4 in the control group.

Table 1. *Cont.*

First Author, Year	AF Population *n* = Sample; Age (Years); Sex (%); AF Subtype	Exercise Intervention *Frequency; Intensity; Time; Type*	Follow-Up and Impact on Health Outcomes (SAE, Physical Capacity, AF Specific Outcomes, and QoL)
Kato, 2019	*n* = 68 Intervention: 67 years 71% male Control: 65 years 90% male Persistent AF	*Frequency*: 1–2 sessions/week, for 24 weeks *Intensity*: Moderate intensity *Time*: 60 min *Type*: Cardiac rehabilitation	Follow-up: 6 months Physical capacity: Significant increases in the 6MWT form 545(123) m to 596 (95) m and also the $V\dot{\,}O_2$-peak from 17.8 (3.4) mL kg^{-1} min^{-1} to 19.8 (4.6) mL kg^{-1} min^{-1} in the intervention group after 6 months with no significant changes in the control group. AF specific: During the six-month follow-up period, 21.4% (6 patients) of the intervention group had AF recurrence and 25.8% (8 patients) in the control group with a risk ratio of 0.83 (95%CI, 0.33 to 2.10). 1 thoracic compression fracture not intervention-related, 1 developed hypothyroidism during the intervention period and no cardiovascular adverse events in either group.
Melo, 2019	*n* = 63 Intervention: 69 years 79% male Control: 66 years 75% male Persistent AF	*Frequency*: 2 sessions/week, for 6 months *Intensity*: Vigorous intensity *Time*: 60 min *Type*: HIIT	Follow-up: 6 months QoL: HQL-14 improved between the groups but with no significant differences. Physical capacity: Significant increases in $V\dot{\,}O_2$-peak after 6 months in both intervention and control; from 12.6 (1.7) mL kg^{-1} min^{-1} to 15.0 (2.3) mL kg^{-1} min^{-1} in intervention and 12.1 (1.7) mL kg^{-1} min^{-1} 15.6 (2.3) mL kg^{-1} min^{-1} in the control.
Wahlström, 2020	*n* = 152 Intervention 65 years 47% male Active control: 63 years 52% male Control: 64 years 49% male Paroxysmal AF	*Frequency*: 1 session/week, for 12 weeks *Intensity*: NR (predict light) *Time*: 60 min/session *Type*: Yoga	Follow-up: 12 weeks QoL: SF-36 significantly improved within Medi-yoga intervention group in the Bodily Pain from 70 (27) to 83 (19), (*p* = 0.014), General Health from 61 (18) to 70 (17), (*p* = 0.037), Social Function from 75 (28) to 88 (18), (*p* = 0.029), Mental Health from 64 (16) to 72 (16), (*p* = 0.030) and Mental Component Summary score from 40 (11) to 46 (9), (*p* = 0.019), subscales, however, no significance between group effects. No adverse events.
Cohort studies			
Ahn, 2021	*n* = 66,692 60 years 64% male Newly diagnosed AF	Cohort stratified by: Persistent non-exercisers (30.5%) New exercisers (17.8%) Exercise dropouts (17.4%) Exercise maintainers (34.2%) *Exercise "Yes" = performing moderate (>30 min) or vigorous intensity exercise (>20 min), at least once a week. Exercise "No" = not engaging in any moderate or vigorous intensity exercise. Therefore, persistent non-exercisers (No to No), new exercisers (No to Yes), exercise dropouts (Yes to No), and exercise maintainers (Yes to Yes).*	Follow-up: 2 years SAE: The new exerciser and exercise maintainer groups were associated with a lower risk of HF compared to the persistent non-exerciser group: the hazard ratios (HRs) (95% Cis) were 0.95 (0.90–0.99) and 0.92 (0.88–0.96), respectively (*p* < 0.001). Performing exercise any time before or after AF diagnosis was associated with a lower risk of mortality compared to persistent non-exercising: the HR (95% CI) was 0.82 (0.73–0.91) for new exercisers, 0.83 (0.74–0.93) for exercise dropouts, and 0.61 (0.55–0.67) for exercise maintainers (*p* < 0.001). For ischemic stroke, HRs were 10%–14% lower in patients of the exercise groups, yet differences were statistically insignificant (*p* = 0.057). Energy expenditure of 1000–1499 MET-min/wk (regular moderate exercise 170–240 min/wk) was consistently associated with a lower risk of each outcome based on a subgroup analysis of the new exerciser group.
Bonnesen, 2021	*n* = 1410 75 years 46% female Participants had AF risk factors but no prior AF diagnosis	A dynamic parameter describing within-individual changes in daily physical activity, i.e., average daily activity in the last week compared to the previous 100 days, was computed and used to model the onset of AF.	Follow-up: 3.5 years A 1 h decrease in average daily physical activity was associated with AF onset the next day (odds ratio 1.24 (1.18–1.31)). This effect was modified by overall level of activity (*p* < 0.001 for interaction), and the signal was strongest in the tertile of participants with lowest activity overall (low: 1.62 (1.41–1.86), mid: 1.27 (1.16–1.39), and high: 1.10 (1.01–1.19)).
Buckley, 2021a	*n* = 23,894 68 years 30% female Included paroxysmal, persistent and permanent AF	Comprehensive cardiac rehabilitation programme *	Follow-up: 18 months Exercise-based CR was associated with 68% lower odds of all-cause mortality (odds ratio, 0.32; 95% CI, 0.29–0.35), 44% lower odds of rehospitalisation (0.56; 95% CI, 0.53–0.59), and 16% lower odds of incident stroke (0.84; 95% CI, 0.72–0.99) compared with propensity-score matched controls. The beneficial association of exercise-based CR on all-cause mortality was independent of sex, older age, comorbidities, and AF subtype.
Buckley, 2021b	*n* = 9808 70 years 32% female Included paroxysmal AF	Comprehensive cardiac rehabilitation programme *	Follow-up: 2 years Progression from paroxysmal AF to sustained AF (persistent/permanent) at 2-year follow-up was proportionally lower with 19.3% (*n* = 617 of 3197 patients) in the exercise-based CR cohort compared to 24.5% (*n* = 909 of 3716 patients) in the matched controls (OR 0.74, 95% CI: 0.66–0.83).

Table 1. *Cont.*

First Author, Year	AF Population *n* = Sample; Age (Years); Sex (%); AF Subtype	Exercise Intervention *Frequency; Intensity; Time; Type*	Follow-Up and Impact on Health Outcomes (SAE, Physical Capacity, AF Specific Outcomes, and QoL)
Garnvik, 2020	*n* = 1117 Inactive: 73 years 61% male Not meeting: 71 years 67% male Meeting: 69 years 78% male All AF subtypes	Cohort stratified into 3 groups: (1). Inactive, reflecting no PA or less than once a week. (2). Below guideline amount, reflecting < 150 min of moderate intensity or <75 min of vigorous intensity/week. (3). At or above guideline amount, ≥150 min moderate or ≥75 min vigorous intensity/week.	Follow-up: up to 9 years or an event Atrial fibrillation patients meeting PA guidelines had lower risk of all-cause (hazard ratio (HR) 0.55, 95% confidence interval (CI) 0.41–0.75) and CVD mortality (HR 0.54, 95% CI 0.34–0.86) compared with inactive patients. The respective HRs for CVD morbidity and stroke were 0.78 (95% CI 0.58–1.04) and 0.70 (95% CI 0.42–1.15). Each 1-MET higher eCRF was associated with a lower risk of all-cause (HR 0.88, 95% CI 0.81–0.95), CVD mortality (HR 0.85, 95% CI 0.76–0.95), and morbidity (HR 0.88, 95% CI 0.82–0.95).
Hegbom, 2006 (Originally an RCT, but results presented as a single arm pre-post study)	*n* = 30 62 years 88% male Permanent AF	*Frequency*: 3 sessions/week for 8 weeks *Intensity*: 70–90% of maximum heart rate *Time*: 1.25 h *Type*: Three 15 min periods of aerobics at HRmax, interrupted by strengthening exercise for the back, thighs, and stomach 15 min of stretching and relaxation	Follow-up: 8 weeks Physical capacity: Increase in cumulated work at Borg-17 scale: The intervention group (39% ± 38%) and control group (42% ± 35%). Quality of life change to follow-up, SF-36 (within intervention group): from 49 ± 6 pre-training to 52 ± 6 post-training ($p < 0.05$).
Lakkireddy, 2013	*n* = 49 61 years 47% male Paroxysmal AF	*Frequency*: 2 sessions/week for 12 weeks *Intensity*: NR (predict light) *Time*: NR *Type*: Yoga	Follow-up: 12 weeks AF specific: Yoga training reduced symptomatic AF episodes (3.8 ± 3 vs. 2.1 ± 2.6, $p < 0.001$), symptomatic non-AF episodes (2.9 ± 3.4 vs. 1.4 ± 2.0; $p < 0.001$), asymptomatic AF episodes (0.12 ± 0.44 vs. 0.04 ± 0.20; $p < 0.001$). QoL: SF-36 significantly improved after the intervention period (Physical Functioning from 85.0 (80.0–95.0) to 90.0 (85.0–95.0), ($p = 0.017$), General Health from 65.0 (50.0–77.5) to 75.0 (65.0–82.5), ($p < 0.001$), Vitality from 84.0 (68.0–88.0) to 91.0 (80.0–95.8), ($p < 0.001$), Social Functioning from 100.0 (75.0–100.0) to 100.0 (90.0–100.0), ($p = 0.019$), and Mental Health from 75.0 (65.0–85.0) to 80.0 (70.0–86.0), ($p < 0.001$). No adverse events.
Younis, 2018	*n* = 304 (pre-post AF arm) 68 years 76% male AF subtype NR	*Frequency*: 2 sessions/week for 6 months *Intensity*: NR *Time*: 60 min *Type*: Cardiac rehabilitation	Follow-up: 6-months Physical capacity: Significant improvement (delta > 5%) was achieved among 194 (64%) patients with AF.

AFEQT, atrial fibrillation quality of life survey; AF, atrial fibrillation; AF-QoL-18, quality of life questionnaire for patients with atrial fibrillation; MET, metabolic equivalent (one MET = 3.5 mL.g.min of oxygen consumption); RCT, randomised controlled trial; SAE, serious adverse event; SF-36, 36-item short form survey; QoL, quality of Life; 6MWT, 6 min walk test. * Individual exercise programme details unknown due to clustering of data from multiple sites using a network research database.

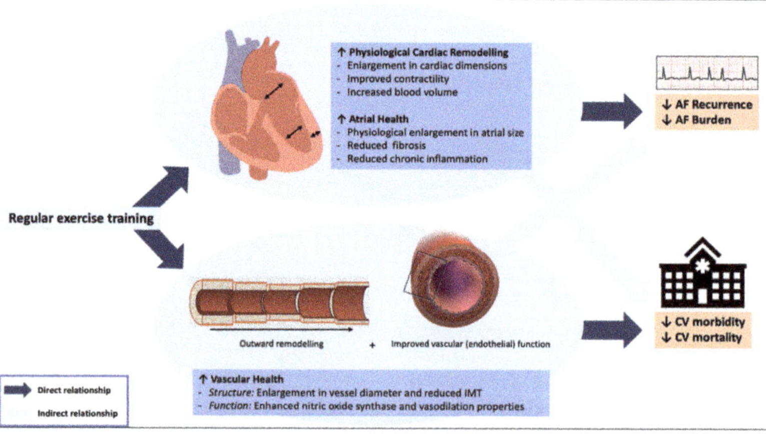

Figure 1. Proposed mechanisms explaining the long-term AF protection from regular exercise training. We propose atrial health has a direct impact on AF recurrence/burden and an indirect impact on AF morbidity whereas vascular health has a direct impact on AF morbidity and an indirect impact on AF recurrence/burden. Dark blue arrows represent a direct relationship. Light blue arrows represent an indirect relationship.

3. Evidence Informing AF-Specific Rehabilitation

3.1. Study Characteristics

Table 1 summarises the reviewed primary research and available data of interest extracted from each study. Of the included 21 studies (22 publications), 13 were randomised controlled trials and 8 were pre-post or cohort studies. Sample size ranged between 66,692 [19] and 18 [20], and female representation ranged between 77% [18] and 12% [21], with 11 studies investigating paroxysmal and/or persistent AF subtypes, [18,22–31], 4 permanent AF, [21,32–34], 2 all-AF subtypes [2,35], and 4 studies did not stratify their results by AF subtype [17,19,36,37]. Included interventions were moderate-intensity continuous training, high-intensity interval training, cardiac rehabilitation (moderate-vigorous intensity continuous training and resistance training), low-to-moderate continuous training (Yoga, Qigong), inspiratory muscle training, and resistance plus moderate-intensity continuous training (circuit training). Follow-up for outcomes of interest (SAE, physical capacity, AF-specific outcomes, and quality of life) ranged between 8-weeks [21] and 9-years [35].

3.2. Intervention/Cohort Details and Long-Term Outcomes

Population-based studies. In a cohort study of 66,692 participants with newly diagnosed AF, performing > 30 min of moderate or >20 min of vigorous intensity exercise at least once per week was associated with significantly lower risk of heart failure and mortality (and a 10–14% lower risk of ischaemic stroke, although the latter was not significant) [19]. In subgroup analyses of patients with AF who went from previously inactive to active, an energy expenditure of 1000–1499 MET-minutes/week, corresponding to 170–240 min per week of moderate intensity exercise, was consistently associated with a lower risk of mortality, stroke, and heart failure in patients with AF [19]. In another cohort study of 1117 patients with AF, participants were stratified by activity level (i.e., inactive, not meeting physical activity guidelines, and meeting physical activity guidelines). The authors reported that meeting the 150 min of moderate-intensity physical activity guidelines was associated with lower risk of all-cause mortality, cardiovascular-related mortality, cardiovascular morbidity, and stroke compared to those not meeting the guidelines and those who were inactive [35]. This work highlights the potential potency of meeting (150 min of moderate intensity physical activity/week) and exceeding the general physical activity guidelines (170–240 min of moderate intensity physical activity/week) on long-term clinical outcomes in patients with AF. However, as these data are from cohort studies, the findings are associations, not causal.

Intervention studies. The most researched exercise-based intervention included in this review was cardiac rehabilitation with seven included studies across four RCTs [22,24,28,29] and three cohort studies [2,30,37]. There were consistent benefits shown in studies that investigated comprehensive cardiac rehabilitation or an intervention designed to reflect cardiac rehabilitation. The various types of intervention included an exercise programme consisting of 1–3 sessions/week of 30–60 min moderate-to-vigorous intensity exercise for 9-weeks to 6-months. A variety of benefits was observed across all studies including fewer major adverse cardiovascular events such as mortality, hospitalisation, and stroke; [2] less progression of AF subtype from paroxysmal to sustained AF; [30] lower AF recurrence; [22] higher physical capacity; [24,28,29,37] and better quality of life [24,28].

A total of four included studies investigated less traditional low-moderate intensity continuous exercise programmes including Yoga, [31] Mediyoga (a therapeutic form of meditative yoga based on deep breathing) [26,27], and Qigong (slow and graceful movements with a focus on breathing) [33]. In a pre-post study, Lakkireddy et al. showed that following 12-weeks of Yoga twice a week, episodes of AF were significantly reduced and quality of life significantly improved (including Physical Functioning, General Health, Vitality, Social Functioning, and Mental Health subscales). More recently, Wahlström et al. demonstrated that 12 weeks of Mediyoga once per week resulted in significantly increased physical capacity (measured via 6MWT) and the mental component of quality of life compared to a control [27]. In a later study, Mediyoga was shown to increase within arm improvements in bodily pain, general health, social function, mental health, and mental component summary scores (as measured by the Short-Form Health Survey) within the MediYoga group, but no significant differences were seen when compared to an active control or usual care [26]. Finally, two 90 min sessions of Qigong for 16-weeks demonstrated increased physical capacity compared to controls (measured via the 6MWT) [33]. This work provides promise for alternative low-moderate intensity continuous exercise, such as Yoga, for improved quality of life and even physical capacity for people with AF. No known study has yet explored whether these benefits also translate to clinical or AF-specific outcomes following this type of exercise.

Three studies investigated the effect of vigorous-intensity exercise for patients with AF. A programme of three sessions per week of vigorous-intensity exercise for 12-weeks was shown to significantly improve time in AF (reduced AF burden), improve physical capacity, and improve quality of life compared to usual care across two RCTs [18,32]. However, in one post hoc analysis from an RCT, two sessions of high-intensity interval exercise per week for 6-months improved quality of life and $\dot{V}O_2$-peak in both intervention and usual care groups (no significant difference between groups). However, the sample of patients in this study underwent cardiac resynchronisation therapy and represented only 18 participants with AF across both groups [20]. One RCT investigated low- vs high-intensity exercise consisting of two sessions of interval and circuit-based training twice per week for 12-weeks [25]. Both groups demonstrated an improvement in $\dot{V}O_2$-peak but no group-time interaction was observed. Furthermore, there was no difference in AF burden (measured as ECGs with AF/ECGs without AF) between the two exercise groups. Although the measurement of AF burden with non-continuous methods (as conducted in the latter study) has its limitations [38], this work highlights a topical argument regarding whether we should focus on exercise duration or exercise intensity.

3.3. Should We Promote Increased Intensity or Duration of Exercise for Improved Outcomes?

By observing data from other more intensively researched cardiovascular conditions [39,40], it is possible that the overall energy expenditure is the most important variable for optimal patient outcomes. In line with this hypothesis, two systematic reviews with meta-analyses and meta-regression have shown that total energy expenditure of an overall exercise programme was the strongest predictor of improvement in exercise capacity for patients with heart failure [39,40]. It was concluded that at least 460 kcal of weekly energy expenditure may elicit the greatest changes in cardiorespiratory fitness for patients with heart failure (regardless of how those kcals are achieved, i.e., high-intensity interval training vs. moderate-intensity continuous training) [39]. Similarly, in a multi-centre comparison, Williams et al. [41] found that in both healthy populations and participants with cardiovascular disease, higher amounts and intensity of exercise increased the likelihood of cardiorespiratory fitness improvement. Furthermore, it has been shown (in a healthy sample) that individual cardiorespiratory 'non-response' to exercise training is abolished by increasing the volume of aerobic continuous exercise [42]. Therefore, it seems that by increasing the overall exercise dose, whether by increasing the volume and/or intensity of an exercise programme, more individuals (even those most likely not to respond) have improved odds of benefiting from the programme. The overall dose of an exercise programme is a product of the frequency of sessions, intensity, and time (duration of individual sessions and programme). All of these can be manipulated to personalize the exercise programme.

By taking an overall exercise dose approach, a *minimum* exercise threshold of 360–720 MET-minutes/week would correspond to 60–120 min of exercise training per week at moderate-to-vigorous intensity (Graphical Abstract). This proposed threshold is typical of a cardiac rehabilitation programme and the majority of aerobic-type exercise interventions included within this review, which have demonstrated clinical benefit. Nonetheless, exceeding this exercise dose would increase one's likelihood of benefit, whether by increasing the number of sessions per week or the intensity of exercise in an individual session. For example, in a rigorous between and within study design, 78 healthy participants were divided into five groups comprising 1–5 60 min exercise sessions for 6-weeks. Non-responders in V'O_2-peak at 6-weeks were prescribed an additional two sessions per week. Findings showed that that in groups 1, 2, 3, 4, and 5, 69%, 40%, 29%, 0%, and 0% of individuals, respectively, were non-responders. Interestingly, after increasing the exercise programme by an additional two sessions for all non-responders across all groups, all non-response in V'O_2-peak was eliminated [42]. This highlights the potency of increasing the dose of an exercise programme via increased exercise volume. However, increasing intensity and duration of exercise may enhance the befits from exercise further. Using a multi-centre comparison of V'O_2-peak trainability between high- and low-volume interval training and moderate-intensity continuous training, with a total of 677 participants representing 18 different intervention studies, high-volume and high-intensity interval training had more responders in V'O_2-peak improvement compared to low-volume and high-intensity interval training and moderate intensity continuous training [41]. Thus, gradually increasing both intensity and duration of exercise may be the optimal strategy for reducing 'non-response' in clinical exercise programmes. This concept could be titrated on an individualised basis for those who may not respond to the initial dose. This may explain the previously discussed findings that increasing one's activity level up to three-fold the current exercise guidelines (1000–1499 MET-minutes/week) had the strongest association with improved major adverse cardiovascular outcomes at 2-year follow-up in previously inactive patients with AF [19]. However, this was based on self-reported data, which can vary considerably from device-based measurement depending on the method used and population investigated [43]. Therefore, this may represent an elevated threshold for improved long-term prognosis in patients with AF. It is also important to note that there has been no research investigating non-response to exercise training in patients with AF, and this warrants future investigation. This may be particularly important for those with concomitant conditions such as AF and heart failure (especially those with preserved ejection fraction) who may be less responsive to initial training [44,45].

3.4. Mechanisms of AF Protection from Long-Term Exercise Training

Although exercise training improves individual risk factors, collectively, only ~50% of the exercise-induced benefit to health can be explained by risk factor improvement [46,47]. There is now a clear body of work that has shown the favourable adaptations seen in vascular structure and function following exercise training [48], which can help to explain this so called "risk factor gap" [49–51]. Figure 1 depicts the hypothesised contribution of risk factor modification, specifically physiological cardiac remodelling (enlargement in cardiac dimension, improved contractility, and increased blood volume), atrial health (physiological increase in atrial size with associated reduction in fibrosis and inflammation), and vascular health (increased diameter size, improved vascular function) to explain the benefit of regular exercise training on AF burden and major adverse events. Here, we propose that the beneficial impact of exercise training on physiological cardiac remodelling [52] and an AF substrate [16] has a direct impact on AF burden and recurrence and an indirect impact on AF morbidity. Conversely, we propose that the beneficial impact of exercise training on vascular structure and function [49,53] has a direct impact on AF morbidity and an indirect impact on AF burden and recurrence (Figure 1).

When referring to increased cardiac remodeling and atrial health (particularly increased atrial size), Figure 1 represents *physiological* hypertrophy/cardiac remodelling as a direct result of exercise training. Physiological hypertrophy/cardiac remodelling is associated with enhanced cardiac function whereas *pathological* hypertrophy/cardiac remodelling is associated with increased fibrosis, cardiac dysfunction, and can be arrhythmogenic. Vascular structure typically refers to the thickness of the arterial wall, measured as the intima-media thickness (IMT), and the diameter of the lumen, measured using vascular ultrasound. Vascular function refers to the ability of the artery to vasodilate and/or vasoconstrict dependent on the stimulus used. Commonly used measures of vascular function include flow-mediated dilatation (largely endothelial dependent) and carotid artery reactivity (largely catecholamine dependent) [48].

3.5. Is There an Acute, AF-Specific, Cardioprotective Effect of Exercise for People with AF?

In a novel prospective cohort study of 1410 participants with risk factors for AF and stroke, Bonnesen et al. [36] created a dynamic parameter describing within-individual changes in daily physical activity, i.e., average daily physical activity in the last week compared to the previous 100-day average. They showed that a 1 h decrease in daily physical activity during the last week increased the odds of AF onset the next day by 24%. This provides the first known data suggesting a possible acute AF-specific protective effect of regular physical activity. Furthermore, the strongest association was observed in the group with the lowest activity overall, where a 1 h reduction in an individual's physical activity increased the odds of AF by 60% for the those in the third lowest activity stratum. Interestingly, the most physically active participants were somewhat protected from transient reductions in physical activity with lower odds of AF following acute physical inactivity. Although these findings are exciting, it is important to emphasise that the analysis stems from a sub-study of a larger prospective study and there are limitations with regard to the detail of physical activity data retrieved from implantable loop recorders. Further, there is the possibility of reverse causality, whereby undetected AF or other condition-specific symptoms may have affected physical activity levels prior to an AF event.

The idea of immediate and acute cardioprotection from exercise is not a new concept and has been previously termed cardiovascular preconditioning or short-term exercise-induced cardioprotection, typically researched in patients with atherosclerotic CVD [54,55]. These short-term exercise-induced benefits stem from pre-clinical evidence that acute exercise has the ability to activate multiple pathways to confer immediate protection against ischaemic events, reduce the severity of potentially lethal ischaemic myocardiac injury, and act as a physiological first line of defence [54]. From an AF-specific perspective, it is possible that the balance between parasympathetic and sympathetic autonomic regulation is involved in the relationship between acute physical (in)activity and AF. For example, AF has been proposed as a disorder of autonomic tone [56] and in 'athletic AF', vagal stimulation has an impact on both the atrial refractory period and action potential duration, both of which are key in the development and progression of AF [16]. Nevertheless, in an AF-specific context at least, the mechanisms of any potential acute cardioprotection following short-term or even a single bout of exercise are unknown and warrant future investigation. Other novel work by Linz et al. [57] reported, in 72 patients undergoing pacemaker monitoring for both AF and respiratory disturbances, that nights with more sleep-disordered breathing conferred a higher risk of subsequent AF. Within each patient, nights with the highest sleep apnoea severity conferred a 1.7-fold increase in the risk of having at least 5 min of AF during the same day, compared with the best rated night of sleep. This provides further insight into the acute effects of lifestyle-related risk factors and their potential impact on the heart. This may even partially explain the highly variable patterns seen in paroxysmal AF. In general, AF detection is more likely if we "*look longer, look harder and look with more sophisticated methods* ... ." [58]. Therefore, more advanced longitudinal monitoring of these dynamic risk factors (e.g., physical activity, sleep quality etc.) may not only help explain the causes of arrythmia but also better inform clinical management and lifestyle interventions beyond traditional risk factors.

4. Limitations

Although work in the area of AF-specific exercise rehabilitation is progressing, there are few, relatively small RCTs that have measured AF-specific outcomes such as AF recurrence, AF burden, and AF-specific quality of life. Moreover, detailed and consistent reports of exercise dose (frequency, intensity, time, and type) can be improved to enhance comparison and help determine the most effective interventions for people with AF. There may be differing effects of exercise training across different AF subtypes (paroxysmal, persistent, and permanent); however, there is currently insufficient evidence to meaningfully stratify results in this manner, and most research to date has included mixed AF subtype cohorts. Although we aimed to propose a more nuanced AF-specific exercise rehabilitation recommendation, this is only the first step towards a personalised exercise prescription for people with AF. The latter necessitating individual views and preferences and the use of the most relevant outcomes to those using the intervention, which is an area requiring further research. It is important to note that interventional research in females is limited, with the majority of patients across studies being male. As there are known sex-based differences in terms of physical activity and incident AF (i.e., females may have a higher upper safe threshold compared to males), which we have previously discussed [16]. Whether this is the case for secondary prevention in patients with AF following an intervention warrants further investigation. In the search for the most effective interventions, participant adherence is critical; therefore, reporting intervention adherence and fidelity as part of a high-quality process evaluation would help to determine real-world feasibility and would be integral to ascertain 'active ingredients' of rehabilitation programmes.

5. Perspectives of Future Directions

Future work in this area should explore whether there are any negative consequences of higher levels of physical activity and exercise training for secondary prevention in patients with AF. In particular, given that at very high levels of vigorous endurance training, such as >5000 MET-minutes/week or 5-to 10 fold the existing physical activity guidelines, we observe an increased prevalence of AF, also known as athletic AF [16]. The exciting prospect of the immediate effects of lifestyle-related risk factors, such as physical activity and sleep quality, warrants further research to understand these dynamic factors and explore strategies to implement such knowledge into personalised treatment strategies. Moreover, suitably powered exercise-based rehabilitation interventions, which include rigorous measures of AF-specific outcomes (e.g., AF burden measured via continuous rhythm monitoring), are needed. The measurement of dynamic risk factors and AF burden is becoming more feasible as new technology and smart devices allow continuous and non-invasive measurement of biomarkers and AF rhythm detection. More research is needed to investigate if there are potentially different therapeutic responses between different AF subtypes. In particular, there is limited research for females and people with permanent AF. Finally, it is central to remember the individual patient in clinical research and the intervention components and outcomes that may be most important to them.

6. Conclusions

We propose that a minimum exercise threshold of 360–720 MET-minutes/week, corresponding to 60–120 min exercise per week at moderate-to-vigorous intensity, could be an evidence-based recommendation for patients with AF to improve AF-specific outcomes, quality of life, and possibly long-term major adverse cardiovascular events. This minimum threshold is typical of a cardiac rehabilitation programme, although we would like to emphasise that the dose could be progressively increased (via duration and/or intensity of exercise) up to and perhaps even beyond 1000–1499 MET-minutes/week. This would increase the likelihood of beneficial outcomes and reduce the number of 'non-responders'. This could be achieved as part of an individualised and progressive rehabilitation programme. In addition, various forms of Yoga seem to have a consistently beneficial effect on the quality of life in patients with AF and would, therefore, seem to be a promising adjunct within an individualised rehabilitation programme.

Highlights

- Four systematic reviews and 21 primary studies were included that investigated exercise training or physical activity for people with AF.
- Included studies represented a range of interventions and participant characteristics including females, males, and all AF subtypes.

- We propose 360–720 MET-minutes/week, corresponding to ~60–120 min of moderate-vigorous intensity exercise per week as an evidence-based recommendation for patients with AF.
- This recommendation could be achieved in a variety of ways to suit the individual i.e., two 60 min or four 30 min sessions of moderate-intensity exercise or three 20 min sessions of high-intensity exercise, for example.
- Further work is needed to explore the potential of immediate effects of lifestyle-related risk factors, such as physical activity and sleep quality, on AF recurrence and burden.
- Furthermore, adequately powered RCTs are needed to investigate the effectiveness of exercise-based rehabilitation on AF-specific outcome measures.

Author Contributions: B.J.R.B. conceived the study. B.J.R.B., S.S.R. and M.B. completed the data extraction. B.J.R.B., S.S.R., M.B., G.Y.H.L. and D.H.J.T. all reviewed and edited the manuscript. All authors have read and agreed to the published version of the manuscript.

Funding: This research received no external funding.

Institutional Review Board Statement: Not applicable.

Informed Consent Statement: Not applicable.

Conflicts of Interest: B.J.R.B. has received research funding from Bristol-Myers Squibb (BMS)/Pfizer. GYHL is a consultant and speaker for BMS/Pfizer, Boehringer Ingelheim, and Daiichi-Sankyo. No fees are received personally.

References

1. Dibben, G.; Faulkner, J.; Oldridge, N.; Rees, K.; Thompson, D.R.; Zwisler, A.-D.; Taylor, R.S. Exercise-based cardiac rehabilitation for coronary heart disease. *Cochrane Database Syst. Rev.* **2021**, *11*, 1465–1858. [CrossRef]
2. Buckley, B.J.R.; Harrison, S.L.; Fazio-Eynullayeva, E.; Underhill, P.; Lane, D.A.; Thijssen, D.H.J.; Lip, G.Y.H. Exercise-Based Cardiac Rehabilitation and All-Cause Mortality Among Patients With Atrial Fibrillation. *J. Am. Heart Assoc.* **2021**, *10*, e020804. [CrossRef]
3. Buckley, B.J.R.; Harrison, S.L.; Fazio-Eynullayeva, E.; Underhill, P.; Sankaranarayanan, R.; Wright, D.J.; Thijssen, D.H.J.; Lip, G.Y.H. Cardiac rehabilitation and all-cause mortality in patients with heart failure: A retrospective cohort study. *Eur. J. Prev. Cardiol.* **2021**, *28* (Suppl. 1), 1704–1710. [CrossRef]
4. Anderson, L.; Taylor, R. Cardiac rehabilitation for people with heart disease: An overview of Cochrane systematic reviews. *Int. J. Cardiol.* **2014**, *177*, 348–361. [CrossRef]
5. Guthold, R.; Stevens, G.A.; Riley, L.M.; Bull, F.C. Worldwide trends in insufficient physical activity from 2001 to 2016: A pooled analysis of 358 population-based surveys with 1.9 million participants. *Lancet Glob. Health* **2018**, *6*, e1077–e1086. [CrossRef]
6. Jeong, S.-W.; Kim, S.-H.; Kang, S.-H.; Kim, H.-J.; Yoon, C.-H.; Youn, T.-J.; Chae, I.-H. Mortality reduction with physical activity in patients with and without cardiovascular disease. *Eur. Heart J.* **2019**, *40*, 3547–3555. [CrossRef] [PubMed]
7. Thyfault, J.P.; Bergouignan, A. Exercise: One size does not fit all. *J. Physiol.* **2020**, *598*, 3819–3820. [CrossRef]
8. Gevaert, A.; Adams, V.; Bahls, M.; Bowen, T.S.; Cornelissen, V.; Dörr, M.; Hansen, D.; Mc Kemps, H.; Leeson, P.; Van Craenenbroeck, E.M.; et al. Towards a personalised approach in exercise-based cardiovascular rehabilitation: How can translational research help? A 'call to action' from the Section on Secondary Prevention and Cardiac Rehabilitation of the European Association of Preventive Cardiology. *Eur. J. Prev. Cardiol.* **2020**, *27*, 1369–1385. [CrossRef] [PubMed]
9. Pelliccia, A.; Sharma, S.; Gati, S.; Bäck, M.; Börjesson, M.; Caselli, S.; Collet, J.-P.; Corrado, D.; Drezner, J.A.; Halle, M.; et al. 2020 ESC Guidelines on sports cardiology and exercise in patients with cardiovascular disease: The Task Force on sports cardiology and exercise in patients with cardiovascular disease of the European Society of Cardiology (ESC). *Eur. Heart J.* **2021**, *42*, 17–96. [CrossRef] [PubMed]
10. Medicine ACoS. *ACSM's Guidelines for Exercise Testing and Prescription*, 9th ed.; Lippincott Williams & Wilkins: Philadelphia, PA, USA, 2013.
11. Hindricks, G.; Potpara, T.; Dagres, N.; Arbelo, E.; Bax, J.J.; Blomström-Lundqvist, C.; Boriani, G.; Castella, M.; Dan, G.-A.; Dilaveris, P.E.; et al. 2020 ESC Guidelines for the diagnosis and management of atrial fibrillation developed in collaboration with the European Association of Cardio-Thoracic Surgery (EACTS): The Task Force for the diagnosis and management of atrial fibrillation of the European Society of Cardiology (ESC) Developed with the special contribution of the European Heart Rhythm Association (EHRA) of the ESC. *Eur. Heart J.* **2021**, *42*, 373–498. [CrossRef] [PubMed]
12. Risom, S.S.; Zwisler, A.-D.; Johansen, P.P.; Sibilitz, K.L.; Lindschou, J.; Gluud, C.; Taylor, R.S.; Svendsen, J.H.; Berg, S.K. Exercise-based cardiac rehabilitation for adults with atrial fibrillation. *Cochrane Database Syst. Rev.* **2017**, *2*, CD011197. [CrossRef]
13. Smart, N.A.; King, N.; Lambert, J.D.; Pearson, M.J.; Campbell, J.; Risom, S.S.; Taylor, R.S. Exercise-based cardiac rehabilitation improves exercise capacity and health-related quality of life in people with atrial fibrillation: A systematic review and meta-analysis of randomised and non-randomised trials. *Open Heart* **2018**, *5*, e000880. [CrossRef]

14. Reed, J.L.; Terada, T.; Chirico, D.; Prince, S.A.; Pipe, A.L. The Effects of Cardiac Rehabilitation in Patients With Atrial Fibrillation: A Systematic Review. *Can. J. Cardiol.* **2018**, *34* (Suppl. 2), S284–S295. [CrossRef]
15. Elkhair, A.A.; Barraclough, D.; Boidin, M.; Buckley, B.; Lane, D.; Lip, G.; Thijssen, D.; Williams, N. Effect of Different Types of Exercise on Quality of Life for Patients with Atrial Fibrillation: Systematic Review. Prospero ID CRD42021231102. Available online: https://www.crd.york.ac.uk/prospero/display_record.php?RecordID=231102 (accessed on 10 February 2022).
16. Buckley, B.; Lip, G.Y.H.; Thijssen, D.H.J. The counterintuitive role of exercise in the prevention and cause of atrial fibrillation. *Am. J. Physiol. Heart Circ. Physiol.* **2020**, *319*, H1051–H1058. [CrossRef]
17. Luo, N.; Merrill, P.; Parikh, K.S.; Whellan, D.J.; Piña, I.L.; Fiuzat, M.; Kraus, W.E.; Kitzman, D.W.; Keteyian, S.J.; O'Connor, C.M.; et al. Exercise Training in Patients With Chronic Heart Failure and Atrial Fibrillation. *J. Am. Coll. Cardiol.* **2017**, *69*, 1683–1691. [CrossRef]
18. Malmo, V.; Nes, B.M.; Amundsen, B.H.; Tjonna, A.-E.; Stoylen, A.; Rossvoll, O.; Wisloff, U.; Loennechen, J.P. Aerobic Interval Training Reduces the Burden of Atrial Fibrillation in the Short Term: A Randomized Trial. *Circulation* **2016**, *133*, 466–473. [CrossRef]
19. Ahn, H.-J.; Lee, S.-R.; Choi, E.-K.; Han, K.-D.; Jung, J.-H.; Lim, J.-H.; Yun, J.-P.; Kwon, S.; Oh, S.; Lip, G.Y.H. Association between exercise habits and stroke, heart failure, and mortality in Korean patients with incident atrial fibrillation: A nationwide population-based cohort study. *PLoS Med.* **2021**, *18*, e1003659. [CrossRef]
20. Melo, X.; Abreu, A.; Santos, V.; Cunha, P.; Oliveira, M.; Pinto, R.; Carmo, M.; Fernhall, B.; Santa-Clara, H. A Post hoc analysis on rhythm and high intensity interval training in cardiac resynchronization therapy. *Scand. Cardiovasc. J.* **2019**, *53*, 197–205. [CrossRef]
21. Hegbom, F.; Stavem, K.; Sire, S.; Heldal, M.; Orning, O.M.; Gjesdal, K. Effects of short-term exercise training on symptoms and quality of life in patients with chronic atrial fibrillation. *Int. J. Cardiol.* **2007**, *116*, 86–92. [CrossRef]
22. Rienstra, M.; Hobbelt, A.H.; Alings, M.; Tijssen, J.G.P.; Smit, M.D.; Brügemann, J.; Geelhoed, B.; Tieleman, R.G.; Hillege, H.L.; Tukkie, R.; et al. Targeted therapy of underlying conditions improves sinus rhythm maintenance in patients with persistent atrial fibrillation: Results of the RACE 3 trial. *Eur. Heart J.* **2018**, *39*, 2987–2996. [CrossRef]
23. Risom, S.S.; Zwisler, A.-D.; Rasmussen, T.B.; Sibilitz, K.L.; Madsen, T.L.; Svendsen, J.H.; Gluud, C.; Lindschou, J.; Winkel, P.; Berg, S.K. Cardiac rehabilitation versus usual care for patients treated with catheter ablation for atrial fibrillation: Results of the randomized CopenHeartRFA trial. *Am. Heart J.* **2016**, *181*, 120–129. [CrossRef]
24. Risom, S.S.; Zwisler, A.-D.; Sibilitz, K.L.; Rasmussen, T.B.; Taylor, R.S.; Thygesen, L.C.; Madsen, T.S.; Svendsen, J.H.; Berg, S.K. Cardiac Rehabilitation for Patients Treated for Atrial Fibrillation With Ablation Has Long-Term Effects: 12-and 24-Month Follow-up Results From the Randomized CopenHeart(RFA) Trial. *Arch. Phys. Med. Rehabil.* **2020**, *101*, 1877–1886. [CrossRef] [PubMed]
25. Skielboe, A.K.; Bandholm, T.Q.; Hakmann, S.; Mourier, M.; Kallemose, T.; Dixen, U. Cardiovascular exercise and burden of arrhythmia in patients with atrial fibrillation—A randomized controlled trial. *PLoS ONE* **2017**, *12*, e0170060. [CrossRef] [PubMed]
26. Wahlström, M.; Rosenqvist, M.; Medin, J.; Walfridsson, U.; Rydell-Karlsson, M. MediYoga as a part of a self-management programme among patients with paroxysmal atrial fibrillation—A randomised study. *Eur. J. Cardiovasc. Nur.* **2020**, *19*, 74–82. [CrossRef] [PubMed]
27. Wahlstrom, M.; Karlsson, M.R.; Medin, J.; Frykman, V. Effects of yoga in patients with paroxysmal atrial fibrillation—A randomized controlled study. *Eur. J. Cardiovasc. Nur.* **2017**, *16*, 57–63. [CrossRef]
28. Joensen, A.M.; Dinesen, P.T.; Svendsen, L.T.; Hoejbjerg, T.K.; Fjerbaek, A.; Andreasen, J.; Sottrup, M.B.; Lundbye-Christensen, S.; Vadmann, H.; Riahi, S. Effect of patient education and physical training on quality of life and physical exercise capacity in patients with paroxysmal or persistent atrial fibrillation: A randomized study. *J. Rehabil. Med.* **2019**, *51*, 442–450. [CrossRef]
29. Kato, M.; Ogano, M.; Mori, Y.; Kochi, K.; Morimoto, D.; Kito, K.; Green, F.N.; Tsukamoto, T.; Kubo, A.; Takagi, H.; et al. Exercise-based cardiac rehabilitation for patients with catheter ablation for persistent atrial fibrillation: A randomized controlled clinical trial. *Eur. J. Prev. Cardiol.* **2019**, *26*, 1931–1940. [CrossRef]
30. Buckley, B.; Harrison, S.; Fazio-Eynullayeva, E.; Underhill, P.; Lane, D.; Thijssen, D.; Lip, G. Association of Exercise-Based Cardiac Rehabilitation with Progression of Paroxysmal to Sustained Atrial Fibrillation. *J. Clin. Med.* **2021**, *10*, 435. [CrossRef]
31. Lakkireddy, D.; Atkins, D.; Pillarisetti, J.; Ryschon, K.; Bommana, S.; Drisko, J.; Vanga, S.; Dawn, B. Effect of yoga on arrhythmia burden, anxiety, depression, and quality of life in paroxysmal atrial fibrillation: The YOGA My Heart Study. *J. Am. Coll. Cardiol.* **2013**, *61*, 1177–1182. [CrossRef]
32. Osbak, P.S.; Mourier, M.; Kjaer, A.; Henriksen, J.H.; Kofoed, K.F.; Jensen, G.B. A randomized study of the effects of exercise training on patients with atrial fibrillation. *Am. Heart J.* **2011**, *162*, 1080–1087. [CrossRef]
33. Pippa, L.; Manzoli, L.; Corti, I.; Congedo, G.; Romanazzi, L.; Parruti, G. Functional capacity after traditional Chinese medicine (qi gong) training in patients with chronic atrial fibrillation: A randomized controlled trial. *Prev. Cardiol.* **2007**, *10*, 22–25. [CrossRef] [PubMed]
34. Zeren, M.; Demir, R.; Yigit, Z.; Gürses, H.N. Effects of inspiratory muscle training on pulmonary function, respiratory muscle strength and functional capacity in patients with atrial fibrillation: A randomized controlled trial. *Clin. Rehabil.* **2016**, *30*, 1165–1174. [CrossRef] [PubMed]
35. Garnvik, L.E.; Malmo, V.; Janszky, I.; Ellekjær, H.; Wisløff, U.; Loennechen, J.P.; Nes, B.M. Physical activity, cardiorespiratory fitness, and cardiovascular outcomes in individuals with atrial fibrillation: The HUNT study. *Eur. Heart J.* **2020**, *41*, 1467–1475. [CrossRef]

36. Bonnesen, M.P.; Frodi, D.M.; Haugan, K.J.; Kronborg, C.; Graff, C.; Højberg, S.; Køber, L.; Krieger, D.; Brandes, A.; Svendsen, J.H.; et al. Day-to-day measurement of physical activity and risk of atrial fibrillation. *Eur. Heart J.* **2021**, *42*, 3979–3988. [CrossRef] [PubMed]

37. Younis, A.; Shaviv, E.; Nof, E.; Israel, A.; Berkovitch, A.; Goldenberg, I.; Glikson, M.; Klempfner, R.; Beinart, R. The role and outcome of cardiac rehabilitation program in patients with atrial fibrillation. *Clin. Cardiol.* **2018**, *41*, 1170–1176. [CrossRef]

38. Chen, L.Y.; Chung, M.K.; Allen, L.A.; Ezekowitz, M.; Furie, K.L.; McCabe, P.; Noseworthy, P.A.; Perez, M.V.; Turakhia, M.P.; American Heart Association Council on Clinical Cardiology; et al. Atrial Fibrillation Burden: Moving Beyond Atrial Fibrillation as a Binary Entity: A Scientific Statement From the American Heart Association. *Circulation* **2018**, *137*, e623–e644. [CrossRef] [PubMed]

39. Ismail, H.; McFarlane, J.R.; Dieberg, G.; Smart, N.A. Exercise training program characteristics and magnitude of change in functional capacity of heart failure patients. *Int. J. Cardiol.* **2014**, *171*, 62–65. [CrossRef] [PubMed]

40. Vromen, T.; Kraal, J.; Kuiper, J.; Spee, R.; Peek, N.; Kemps, H. The influence of training characteristics on the effect of aerobic exercise training in patients with chronic heart failure: A meta-regression analysis. *Int. J. Cardiol.* **2016**, *208*, 120–127. [CrossRef]

41. Williams, C.J.; Gurd, B.J.; Bonafiglia, J.T.; Voisin, S.; Li, Z.; Harvey, N.; Croci, I.; Taylor, J.L.; Gajanand, T.; Ramos, J.S.; et al. A Multi-Center Comparison of O(2peak) Trainability Between Interval Training and Moderate Intensity Continuous Training. *Front. Physiol.* **2019**, *10*, 19. [CrossRef]

42. Montero, D.; Lundby, C. Refuting the myth of non-response to exercise training: 'non-responders' do respond to higher dose of training. *J. Physiol.* **2017**, *595*, 3377–3387. [CrossRef]

43. Prince, S.A.; Adamo, K.B.; Hamel, M.E.; Hardt, J.; Gorber, S.C.; Tremblay, M. A comparison of direct versus self-report measures for assessing physical activity in adults: A systematic review. *Int. J. Behav. Nutr. Phys. Act.* **2008**, *5*, 56. [CrossRef] [PubMed]

44. Mueller, S.; Winzer, E.B.; Duvinage, A.; Gevaert, A.B.; Edelmann, F.; Haller, B.; Pieske-Kraigher, E.; Beckers, P.; Bobenko, A.; Hommel, J.; et al. Effect of High-Intensity Interval Training, Moderate Continuous Training, or Guideline-Based Physical Activity Advice on Peak Oxygen Consumption in Patients With Heart Failure With Preserved Ejection Fraction: A Randomized Clinical Trial. *JAMA* **2021**, *325*, 542–551. [CrossRef] [PubMed]

45. Buckley, B.J.R.; Lip, G.Y.H.; Thijssen, D.H.J. Effect of Training on Peak Oxygen Consumption in Patients with Heart Failure with Preserved Ejection Fraction. *JAMA* **2021**, *326*, 770–771. [CrossRef] [PubMed]

46. Taylor, R.S.; Unal, B.; Critchley, J.; Capewell, S. Mortality reductions in patients receiving exercise-based cardiac rehabilitation: How much can be attributed to cardiovascular risk factor improvements? *Eur. J. Cardiovasc. Prev. Rehabil.* **2006**, *13*, 369–374. [CrossRef] [PubMed]

47. Mora, S.; Cook, N.; Buring, J.E.; Ridker, P.M.; Lee, I.-M. Physical activity and reduced risk of cardiovascular events: Potential mediating mechanisms. *Circulation* **2007**, *116*, 2110–2118. [CrossRef] [PubMed]

48. Buckley, B.J.R.; Boidin, M.; Thijssen, D.H.J. Assessment of Peripheral Blood Flow and Vascular Function. In *Sport and Exercise Physiology Testing Guidelines*; Taylor & Francis Group: Abingdon, UK, 2022.

49. Green, D.J.; Hopman, M.T.E.; Padilla, J.; Laughlin, M.H.; Thijssen, D.H.J. Vascular Adaptation to Exercise in Humans: Role of Hemodynamic Stimuli. *Physiol. Rev.* **2017**, *97*, 495–528. [CrossRef] [PubMed]

50. Green, D.J.; O'Driscoll, G.; Joyner, M.J.; Cable, N.T. Exercise and cardiovascular risk reduction: Time to update the rationale for exercise? *J. Appl. Physiol.* **2008**, *105*, 766–768. [CrossRef]

51. Thijssen, D.H.J.; Carter, S.; Green, D.J. Arterial structure and function in vascular ageing: Are you as old as your arteries? *J. Physiol.* **2016**, *594*, 2275–2284. [CrossRef]

52. Seo, D.Y.; Kwak, H.-B.; Kim, A.H.; Park, S.H.; Heo, J.W.; Kim, H.K.; Ko, J.R.; Lee, S.J.; Bang, H.S.; Sim, J.W.; et al. Cardiac adaptation to exercise training in health and disease. *Pflug. Arch. Eur. J. Physiol.* **2020**, *472*, 155–168. [CrossRef]

53. Qin, S.; Boidin, M.; Buckley, B.J.R.; Lip, G.Y.H.; Thijssen, D.H.J. Endothelial dysfunction and vascular maladaptation in atrial fibrillation. *Eur. J. Clin. Investig.* **2021**, *51*, e13477. [CrossRef]

54. Thijssen, D.H.J.; Redington, A.; George, K.P.; Hopman, M.T.E.; Jones, H. Association of Exercise Preconditioning With Immediate Cardioprotection: A Review. *JAMA Cardiol.* **2018**, *3*, 169–176. [CrossRef]

55. Thijssen, D.H.J.; Uthman, L.; Somani, Y.; van Royen, N. Short term exercise-induced protection of cardiovascular function and health: Why and how fast does the heart benefit from exercise? *J. Physiol.* **2022**, *600*, 1339–1355. [CrossRef]

56. Coumel, P. Paroxysmal atrial fibrillation: A disorder of autonomic tone? *Eur. Heart J.* **1994**, *15* (Suppl. A), 9–16. [CrossRef]

57. Linz, D.; Brooks, A.G.; Elliott, A.D.; Nalliah, C.J.; Hendriks, J.; Middeldorp, M.; Gallagher, C.; Mahajan, R.; Kalman, J.M.; McEvoy, R.D.; et al. Variability of Sleep Apnea Severity and Risk of Atrial Fibrillation: The VARIOSA-AF Study. *JACC Clin. Electrophysiol.* **2019**, *5*, 692–701. [CrossRef]

58. Lau, Y.C.; Lane, D.A.; Lip, G.Y. Atrial fibrillation in cryptogenic stroke: Look harder, look longer, but just keep looking. *Stroke A J. Cereb. Circ.* **2014**, *45*, 3184–3185. [CrossRef]

Journal of
*Personalized
Medicine*

Review

Current Therapeutic Approach to Atrial Fibrillation in Patients with Congenital Hemophilia

Minerva Codruta Badescu [1,2], Oana Viola Badulescu [3,4], Lacramioara Ionela Butnariu [5,*], Mariana Floria [1,6,*], Manuela Ciocoiu [3], Irina-Iuliana Costache [1,7], Diana Popescu [1], Ioana Bratoiu [8], Oana Nicoleta Buliga-Finis [1] and Ciprian Rezus [1,2]

1 Department of Internal Medicine, "Grigore T. Popa" University of Medicine and Pharmacy, 700115 Iasi, Romania; minerva.badescu@umfiasi.ro (M.C.B.); irina.costache@umfiasi.ro (I.-I.C.); dr.popescu.diana@gmail.com (D.P.); oana_finish@yahoo.com (O.N.B.-F.); ciprian.rezus@umfiasi.ro (C.R.)
2 III Internal Medicine Clinic, "St. Spiridon" County Emergency Clinical Hospital, 700111 Iasi, Romania
3 Department of Pathophysiology, "Grigore T. Popa" University of Medicine and Pharmacy, 700115 Iasi, Romania; oana.badulescu@umfiasi.ro (O.V.B.); manuela.ciocoiu@umfiasi.ro (M.C.)
4 Hematology Clinic, "St. Spiridon" County Emergency Clinical Hospital, 700111 Iasi, Romania
5 Department of Mother and Child Medicine, "Grigore T. Popa" University of Medicine and Pharmacy, 700115 Iasi, Romania
6 Internal Medicine Clinic, "Dr. Iacob Czihac" Emergency Military Clinical Hospital Iasi, 700483 Iasi, Romania
7 Cardiology Clinic, "St. Spiridon" County Emergency Clinical Hospital, 700111 Iasi, Romania
8 Department of Rheumatology and Physiotherapy, "Grigore T. Popa" University of Medicine and Pharmacy, 700115 Iasi, Romania; ioana.bratoiu@umfiasi.ro
* Correspondence: ionela.butnariu@umfiasi.ro (L.I.B.); floria_mariana@yahoo.com (M.F.)

Citation: Badescu, M.C.; Badulescu, O.V.; Butnariu, L.I.; Floria, M.; Ciocoiu, M.; Costache, I.-I.; Popescu, D.; Bratoiu, I.; Buliga-Finis, O.N.; Rezus, C. Current Therapeutic Approach to Atrial Fibrillation in Patients with Congenital Hemophilia. *J. Pers. Med.* **2022**, *12*, 519. https://doi.org/10.3390/jpm12040519

Academic Editor: José Miguel Rivera-Caravaca

Received: 14 February 2022
Accepted: 21 March 2022
Published: 23 March 2022

Abstract: Cardiovascular disease in hemophiliacs has an increasing prevalence due to the aging of this population. Hemophiliacs are perceived as having a high bleeding risk due to the coagulation factor VIII/IX deficiency, but it is currently acknowledged that they also have an important ischemic risk. The treatment of atrial fibrillation (AF) is particularly challenging since it usually requires anticoagulant treatment. The CHA$_2$DS$_2$-VASc score is used to estimate the risk of stroke and peripheral embolism, and along with the severity of hemophilia, guide the therapeutic strategy. Our work provides the most complete, structured, and updated analysis of the current therapeutic approach of AF in hemophiliacs, emphasizing that there is a growing interest in therapeutic strategies that allow for short-term anticoagulant therapy. Catheter ablation and left atrial appendage occlusion have proven to be efficient and safe procedures in hemophiliacs, if appropriate replacement therapy can be provided.

Keywords: atrial fibrillation; anticoagulant; catheter ablation; left atrial appendage occlusion; cardiovascular disease; hemophilia

1. Introduction

Hemophilia is a rare genetic disorder. It is caused by inherited X-linked mutations in the genes encoding coagulation factor VIII (hemophilia A) or factor IX (hemophilia B). It is characterized by a lifelong increased tendency to bleed, which is generally proportional to the degree of coagulation factor deficiency. Of the two types, hemophilia A (HA) is six times more prevalent than hemophilia B (HB) [1]. While women are carriers, men clinically express the disease.

Due to advances in treating hemophilia, the life expectancy of hemophiliacs has currently increased, approaching that of the general male population [2–6]. Once uncommon, age-related diseases have emerged in hemophiliacs and the number of cases is growing. Among the most frequent non-hemophilia-related medical comorbidities are cancer and cardiovascular disease, both associated with old age. Their diagnosis and treatment are extremely challenging because it often requires procedures and drugs that may worsen the already deficient hemostasis. The treatment of cardiovascular diseases in patients from the general population usually includes antithrombotic therapy, either anticoagulant, antiplatelet agents, or both. Evidence-based guidelines for the management of cardiovascular diseases in hemophiliacs are lacking and the antithrombotic treatment is guided by the consensus of experts based on the international guidelines dedicated to non-hemophilia patients [1,7–9].

Atrial fibrillation (AF) is the most common sustained cardiac arrhythmia in the general adult population. The prevalence is 2–4% and rises due to the increase in the number of elderly people worldwide and the intensification of the effort to diagnose AF [10–12]. Observational studies note similar rates of AF in patients with hemophilia compared to the general population [13,14]. Since its treatment usually involves anticoagulant therapy and interventional techniques, AF poses unique therapeutic challenges in hemophiliacs because these patients have a bleeding-prone hemostatic balance. Our work provides an in-depth analysis of the modern therapeutic approach to AF in hemophiliacs, taking into account the new advances in the field. In order to extract the necessary data, we performed an extensive search in the Web of Science. The keywords "atrial fibrillation" and ("hemophilia" or "haemophilia") were searched in the titles and abstracts of the articles. As this review aimed to provide information on the current treatment of atrial fibrillation, the search was restricted to the last 10 years. We manually screened the titles and abstracts of all 138 articles retrieved from the automatic search, removing duplicates and articles in languages other than English.

2. AF Epidemiology and Risk Factors

Age and many of the comorbidities associated with aging are important risk factors for AF [12,15–27]. Current evidence shows that hypertension, diabetes mellitus, dyslipidemia, obesity, and smoking are very common among hemophiliacs [28–32]. Moreover, of these cardiovascular risk factors, hypertension is more prevalent in hemophiliacs than in the general male population [29–33]. The clustering of cardiovascular risk factors is also present, with more than a third of hemophiliacs having metabolic syndrome [33]. The unfavorable cardiovascular risk profile sets in earlier in life in hemophiliacs than in the general population [29,31,34] and the cardiovascular risk factor burden is present, irrespective of hemophilia severity [32,35].

In men, the risk of AF increases steeply after the age of 50 [36]. Thus, with the prolongation of the lifespan of hemophiliacs and the accumulation of risk factors for AF, the number of patients with this arrhythmia is expected to increase. In the largest European cohort of hemophiliacs analyzed to date, the prevalence of AF was 0.84% [13]. A low value of 0.2% was found in patients younger than 60 years of age and seventeen times higher (3.4%) in those over 60 years of age, similar to that of the general population [9]. While there was a very important age-dependent upward trend, the prevalence of AF was inversely correlated with the severity of hemophilia. Other European studies reported a prevalence of 2.2–2.4% [37,38]. The lowest prevalence of AF was found in Asian hemophiliacs (0.28%) [30], and the highest in two North American cohorts, 5.4% in young hemophiliacs [39] and 7.5% in those aged between 54 and 73 [32]. This highlights the negative impact of aging on the prevalence of AF.

3. Assessment of the Ischemic Risk

Blood stasis in the left atrium and impaired contractility of the left atrial appendage are the main contributors to cardioembolism [40]. It is considered that 20–30% of all ischemic strokes and 10% of cryptogenic strokes are the consequence of AF [12]. Generally, AF multiplies the risk of stroke by five [41]. This risk is not homogeneous in the general population, but is dependent on comorbidities and ranges between 1% and 20% [42,43]. Moreover, strokes associated with AF are usually severe, highly recurrent, often fatal, or with permanent disability [12].

The risk of AF-related stroke is influenced by additional factors, therefore, risk scores have been developed to stratify patients and guide the anticoagulant treatment [44–46]. The CHADS$_2$ score was originally used [47], but the CHA$_2$DS$_2$-VASc score outperformed the CHADS$_2$ score and simplified the anticoagulation decision-making as it could identify individuals with a truly low thromboembolic risk that do not require anticoagulant treatment (0 in men and 1 in women) [48,49].

Data on the risk of ischemic stroke in hemophiliacs compared to men in the general population are scarce. Still, the rate of ischemic stroke is expected to be low [50]. A retrospective study showed that the prevalence of ischemic stroke was lower in hemophiliacs with severe disease (0%) than in the general male population (1.5%), but this difference was not statistically significant in hemophiliacs with mild-moderate disease [37]. Of four patients with ischemic stroke, the event was correlated with the presence of AF in one.

Few data are available on the risk of ischemic stroke and systemic embolism in hemophiliacs with AF. CHADS$_2$ and subsequently CHA$_2$DS$_2$-VASc scores were used to estimate this risk, but none have been prospectively validated in a population of hemophiliacs thus far. The CHA$_2$DS$_2$-VASc score was calculated in 33 patients with AF from a large pan-European cohort of almost 4000 hemophiliacs [13]. The score ranged from 0 to 4, driven principally by age and hypertension. The score's mean value was 1.3. Of the 33 hemophiliacs with AF, 16 (48%) had hypertension, three (9%) peripheral vascular disease, three (9%) diabetes, and two (6%) had previous stroke. None had congestive heart failure. Eleven patients (33%) were between 65 and 74 years and three (9%) were 75 years or older. There were ten hemophiliacs with a CHA$_2$DS$_2$-VASc score of 0. Based on the CHA$_2$DS$_2$-VASc score, in hemophiliacs with AF in this cohort, the risk of stroke was low [13].

The French registry included 68 hemophiliacs requiring antithrombotic treatment for acute coronary syndromes (ACS)/coronary artery disease (CAD), AF, or both [51]. Seventeen patients only had AF and one patient had both AF and CAD. The CHA$_2$DS$_2$-VASc score ranged from one to seven. Sixteen patients had a CHA$_2$DS$_2$-VASc score ≥ 2. The score's mean value was three, much higher than previously reported [13]. This difference was not determined by the age of the enrolled patients, but by the load of comorbidities.

Data also came from a short series of cases and case reports. In seven HA patients, the CHA$_2$DS$_2$-VASc score ranged between one and six, with a mean value of three, where age and hypertension were the most important contributors [52]. The patient with severe hemophilia had the highest CHA$_2$DS$_2$-VASc score mainly due to advanced age, previous stroke, and myocardial infarction. Cheung et al. reported the case of a HA patient with mild disease who also had a score of six due to age, left ventricular dysfunction, hypertension, history of stroke, and atherosclerotic carotid stenosis [53]. A very high CHA$_2$DS$_2$-VASc score of seven was reported in a 79-year-old patient with mild HA, based on age, chronic heart failure, hypertension, transient ischemic attack, and coronary artery disease [54].

Although the value of the CHA$_2$DS$_2$-VASc score varies widely in this population, it has become evident that due to the multiple risk factors that accumulate with age, hemophiliacs with AF have an increased risk of ischemic stroke and systemic embolism, regardless of the severity of the disease. Therefore, stroke prevention measures should be discussed with the patient and implemented by a cardiologist–hematologist team.

The first attempt to incorporate a stroke risk score in anticoagulation decision-making was based on the CHADS$_2$ score [55]. Currently, the CHA$_2$DS$_2$-VASc score is endorsed by the ADVANCE (age-related developments and comorbidities in hemophilia) Working Group [8]. The proposed algorithm incorporates both the CHA$_2$DS$_2$-VASc score and the severity of hemophilia. Still, caution is advised as the absolute risk of stroke may be overestimated because the CHA$_2$DS$_2$-VASc score includes parameters that are either irrelevant—female sex—or rarely relevant— peripheral arterial disease and very old age—for the hemophilia population. It has been proposed that only hemophiliacs with a CHA$_2$DS$_2$-VASc score ≥ 3 should be considered with a high risk of stroke [50]. Considering the severity of hemophilia and the risk of bleeding, other researchers have considered that a CHA_2DS_2-VASc score ≥ 2 should be used to start anticoagulation in patients with factor level $\geq 20\%$ and a higher threshold—CHA_2DS_2-VASc score ≥ 4—was suggested in hemophiliacs with a factor level $<20\%$ [8,9].

4. Assessment of the Bleeding Risk

The HAS-BLED score is used in the general AF population for estimating the basal bleeding risk before and during the anticoagulant treatment [56]. A HAS-BLED score ≥ 3 indicates a high risk of bleeding, but this value should not be used as an absolute parameter for withholding or withdrawing anticoagulation. This score must be seen as a tool to reduce the bleeding risk by identifying those risk factors that can be avoided or reversed.

The reliability of this score in hemophiliacs has long been questioned because two major determinants of the risk of bleeding in hemophiliacs—the severity of the disease and the presence of inhibitors—are not directly assessed by this score.

The frequency of bleedings depends on the severity of hemophilia. The highest risk is in patients with a factor level $<1\%$ of normal, who may bleed spontaneously or after minimal trauma [57,58]. Hemophiliacs have a particular hemorrhagic profile. Bleedings are usually recurrent and mainly affect the joints and muscles. Although it is a rare complication, intracranial hemorrhage (ICH) is the most feared because it is life-threatening. Its incidence is higher in hemophiliacs than in the general population, especially in those with severe disease [59,60]. The implementation of modern therapies has led to a major reduction in ICH incidence so that currently, the patients on prophylactic treatment no longer develop ICH [59–62]. Moreover, hypertension has long been associated with ICH [63]. Since hypertension is very prevalent in hemophiliacs, its diagnosis and intensive treatment should be promptly implemented to reduce the ICH risk in the entire hemophiliac population.

Hemophilia patients without inhibitors typically have a predictable response to clotting factor replacement, and bleeding can be reliably prevented or treated. Those with baseline factor levels $>5\%$ and those with severe hemophilia under clotting factor prophylaxis have the lowest risk of bleeding [7]. Patients with inhibitors have a much less predictable hemostatic response to bypassing agents. The problem with developing neutralizing anti-FVIII antibodies is of great importance as it occurs in 25–33% of HA patients [64,65]. Studies have found that 10–20% of bleedings in hemophiliacs with inhibitors are either unresponsive or only partially responsive to bypassing agents [66]. Therefore, the risk of uncontrolled bleeding almost always outweighs the benefit of anticoagulation.

When reported, often in a small series of cases and case reports, the HAS-BLED score was frequently three. In seven HA patients, the HAS-BLED score ranged between one and three, with a mean value of three. Higher values of the score such as five or six have also been reported [54,67].

Until 2021, when the results of the French registry were published, there was a general agreement that the HAS-BLED score underestimated the bleeding risk in hemophiliacs and was considered not suitable for use in this population [50,68]. The French registry provided evidence to the contrary. The HAS-BLED score of the 18 hemophiliacs with AF ranged from zero to four, with a mean value of two [51]. A HAS-BLED score ≥ 3 was associated with increased bleeding risk. In the two years of follow-up under antithrombotic treatment, five out of eight patients with a HAS-BLED score ≥ 3 reported bleeding episodes, while no hemorrhagic events were reported in any of the 10 patients with a HAS-BLED score <3. Of note, none of the hemophiliacs had inhibitors. The median coagulation factor level and the proportion of patients on prophylaxis were similar between the two groups. This was the first report to show that the HAS-BLED score is applicable to hemophiliacs.

The type, intensity, and duration of antithrombotic therapy are determined, on one hand, by the ischemic risk, and on the other hand, by the risk of bleeding and need for replacement therapy, which must ensure a level of coagulation factor that allows for safe antithrombotic treatment. Intensive replacement therapy as required by anticoagulants is unlikely to be sustainable in the long-term. In this context, finding alternative therapeutic solutions has become a priority. Of particular interest are the catheter ablation techniques and the devices for closing the left atrial appendage.

5. Rhythm or Rate Control in AF

When choosing between rhythm and rate control, it should be considered that over time, AF becomes less responsive to treatment or is irreversible [69] and that AF progression leads to an increased risk of ischemic stroke or systemic embolism, a high risk of hospitalization for heart failure [70] and a decrease in the quality of life [71]. The current guideline highlights the fact that the primary indication for rhythm control is to reduce the symptom burden related to the arrhythmia and to improve the quality of life, especially when factors favoring rhythm control are prevailing [12]. While no major differences were observed in cardiovascular mortality or stroke rate [72], the rhythm control strategy improved the left ventricular function and quality of life in patients with heart failure [73].

The drugs used for rate or rhythm control and the antiarrhythmic selection algorithm in hemophiliacs mirror the guidelines for AF management in the general population [12]. One option to achieve rhythm control is through cardioversion and the currently available data in hemophiliacs reflect a preference for the pharmacologic approach [74–78]. The anticoagulant therapy should be started as soon as possible before cardioversion, and concomitant coagulation factor replacement therapy should be administered when necessary. The algorithm that includes transesophageal echocardiography (TEE) should be considered first—when available—because TEE can rule out atrial thrombi and the 3-week period of anticoagulation prior to procedure is avoided [12]. Of note, successful cardioversion does not imply that long-term anticoagulant treatment is no longer necessary. The stroke occurrence in high-risk patients is the same, regardless of the rhythm or rate control strategy adopted [79], therefore, the decision regarding long-term oral anticoagulant treatment should be driven by the presence of the stroke risk factors [12]. Thus, other therapeutic options should be considered in hemophiliacs with a CHA_2DS_2-VASc score of one or higher in whom long-term anticoagulant therapy treatment is not feasible.

Patients with AF require both antiarrhythmic and anticoagulant treatment, therefore, the guidelines provide specific recommendations to help in the selection of the appropriate drug and dose [80]. Recent data on clinical implications of drug–drug interactions show that in patients over 66 years old on anticoagulant treatment with a NOAC —apixaban, dabigatran, rivaroxaban—the bleeding risk was not increased by the concomitant use of amiodarone, diltiazem, and verapamil [81]. If dronedarone is used, it should be noted that it increases the frequency of gastrointestinal bleedings when associated with dabigatran and rivaroxaban and the risk of overall bleeding in patients receiving rivaroxaban [82]. Thus, when selecting AF treatment for hemophiliacs, it should be considered that the underlying hemorrhagic risk will be increased by the administration of the antithrombotic therapy and that some antiarrhythmic drugs may further potentiate this risk.

Due to the increased bleeding risk of hemophiliacs, there is a permanent pursuit for the use of the minimum effective dose of anticoagulant. The increasing availability of thrombin generation assays could facilitate this approach. While conventional coagulation tests provide only partial information on hemostasis, the global coagulation tests evaluate the functionality of all its components. By assessing the dynamics of clot formation, clot resistance, and stability, global coagulation tests reflect the interaction between procoagulants, natural anticoagulants, platelets, and the fibrinolytic system. Guided by a thrombin generation assay, a reduced dose of low molecular weight heparin was safely administered to a HA patient before cardioversion [68].

6. Surgical and Catheter Ablation of AF

AF catheter ablation for pulmonary vein isolation (PVI) is an efficient and safe rhythm control strategy. It represents a first-line therapy in selected patients with symptomatic AF and it is also recommended after antiarrhythmic drug therapy failure [12,83,84]. The interest in catheter ablation is growing because this therapy is perceived as a way to avoid long-term treatment with oral anticoagulants. It was hypothesized that after successful ablation, the risk of thromboembolic events in patients with AF will be the same as in patients without AF [85–87], thus the anticoagulant treatment could be stopped. However, discontinuation of oral anticoagulant therapy is associated with a low risk of stroke only in patients with a low CHA_2DS_2-VASc score [84]. Patients with a previous stroke continue to have a high risk of thromboembolic events despite the successful procedure [88].

Anticoagulant treatment is necessary during the procedure to prevent thrombus formation on sheaths and catheters and at the sites of ablation. The standard protocol using unfractionated heparin is also implemented in hemophiliacs given that during the procedure, the coagulation is restored to normal by the factor replacement therapy [78,84]. Most centers adopt the 300–350 s interval [78], but a slightly lower activated clotting time (ACT) target has been used [77]. There is evidence that even an ACT target range of 225–250 s is safe and effective in specified settings [89,90].

AF ablation is followed by a period of high embolic risk due to atrial stunning, tissue damage caused by the procedure, and possible early recurrence of arrhythmia [77]. Therefore, the anticoagulant treatment is recommended for at least two months after the procedure [12]. Because the predictive value of the CHA_2DS_2-VASc score is maintained after AF ablation regardless of the arrhythmic outcome [88], it is recommended that the decision to stop or continue anticoagulation beyond two months be made based on the patient's risk profile and not on the result of the ablation procedure [12,84].

Cardiac ablation includes the catheter, surgical, and hybrid surgical–catheter ablation. Data on protocol and outcome of the catheter ablation procedure for AF treatment in hemophiliacs are very recent (Table 1). Cryoablation was successfully performed in a HA patient with symptomatic AF that could not be amended through multiple pharmacological attempts [77]. Anticoagulant and replacement factor therapy were used during the PVI procedure, but no long-term antithrombotic therapy was recommended due to the patient's CHA_2DS_2-VASc score of 0.

A total of seven PVI procedures for symptomatic AF—with an average duration of five years—were performed in five HA patients [78]. Three patients remained in sinus rhythm during follow-up, while in two cases, the procedure was repeated at nine and 16 months after the first PVI due to AF recurrence. The success rate of PVI was comparable to that reported in the general population [91] and the complications were only hematoma at the femoral vein puncture site. Procedures were performed under anticoagulant treatment and replacement factor therapy, aiming at a peak coagulation factor level of 80–100% during ablation and in the following 24 h. On the first day after the procedure, the trough level of the coagulation factor was kept above 50%. For at least four weeks after the procedure, patients received either dabigatran 110 mg bid or VKA and bridging with low-molecular-weight heparin (LMWH) until INR >2. A coagulation factor level of at least 20% was considered safe during the anticoagulant treatment.

Surgical AF ablation performed concomitantly with cardiac surgery is an opportunity to add supplementary benefits [92]. A 50-year-old HA patient with mild disease underwent minimally invasive mitral valve repair for valve leaflets prolapse with severe mitral regurgitation. Exclusion of the left atrial appendage and left and right atrial cryoablation for concurrent AF was also performed [93]. His CHA_2DS_2-VASc score was one. At a 2-year follow-up, the patient had sinus rhythm without AF paroxysms. No antithrombotic was recommended after surgery.

Severe complications after catheter ablation occur in 5–7% of patients in the general population, of which 2–3% are life-threatening. Stroke (<1%) and pericardial tamponade (1–2%) are the most severe. Clinical relevant hematoma at the site of venous puncture occurs in 0.9% of patients in the general population. Periprocedural inguinal bleeding has been reported in hemophiliacs, in two cases with a significant decrease in the hemoglobin level [78]. No fatalities have been reported in hemophiliacs to date.

Table 1. Reports on the non-pharmacological treatment of AF.

Author, Year	Patient Age, Sex	Type and Severity of Hemophilia (Baseline Factor Activity Level)	CHA$_2$DS$_2$-VASc Score/HAS-BLED Score	Comorbidities	Procedure/Device	Antithrombotic Treatment after the Procedure and on Long-Term	Coagulation Factor Replacement Treatment	Outcome
Lin et al., 2015 [77]	54 y, M	HA, mild (5%)	0/NR	Obstructive sleep apnea, hemarthroses of peripheral joints	Catheter ablation (PVI)	No antithrombotic treatment	FVIII level 131% before the procedure	No periprocedural complications; No complications at 2-year follow-up
van der Valk et al., 2019 [78]	70 y, NR	HA, mild (35%)	3/NR	NR	Catheter ablation (PVI) Catheter ablation (PVI) -repeated	VKA 3 mo Dabigatran 110 mg bd	FVIII level ≥80% for the procedure and for the first 24 h; FVIII level ≥20% while on anticoagulant (replacement therapy was given when needed)	Groin bleeding with severe anemia (day 5 after the first procedure)
	72 y, NR	HA, severe (<1%)	1/NR		Catheter ablation (PVI) Catheter ablation (PVI) -repeated	VKA 1 mo VKA 1 mo, SAPT with aspirin 2 mo		Groin bleeding with severe anemia (day 3 after the first procedure) Oozing during 4 h at puncture site at the second procedure
	59 y, NR	HA, mild (23%)	0/NR		Catheter ablation (PVI)	Dabigatran 110 mg bd 6 mo		No periprocedural complications
	50 y, NR	HA, severe (<1%)	0/NR		Catheter ablation (PVI)	VKA 6 wk		No periprocedural complications
	55 y, NR	HA, mild (6%)	0/NR		Catheter ablation (PVI)	Dabigatran 110 mg bd 6 wk		No periprocedural complications
Bogachev-Prokophiev et al., 2020 [93]	50 y, M	HA, severe (<1%)	1/NR	Both mitral valve leaflets prolapse with severe regurgitation, recurrent hemarthrosis with limited mobility in the elbow and knee joints	Left and right atrial ablation; left atrial appendage was excluded	No antithrombotic treatment	FVIII level 109% before the procedure	Moderate HF and supraventricular tachycardia during hospitalization; Class I NYHA HF at 2-year follow-up
Cheung et al., 2013 [53]	73 y, M,	HA, mild (8%)	6/NR	CABG, stroke, 90% stenosis of the right ICA from calcified plaque amended by endarterectomy, 50–70% stenosis of the left ICA, hypertension, moderate left ventricular impairment, hyperc-holesterolemia, hepatitis C, COPD	Amplatzer Cardiac Plug	DAPT with aspirin + clopidogrel 6w; SAPT with clopidogrel lifelong	FVIII level 100% for the procedure, ≥80% 3 days ≥30% on DAPT	No periprocedural complications; No thrombotic or coronary events and no bleeding complications at 9-month follow-up
Bhatti et al., 2019 [94]	60 y, F	HB, mild (15%)	3/NR	Sick sinus syndrome status post pacemaker implantation, TIA	Watchman PVI	VKA for 1 mo	FVIII level ≥ 30% on VKA	No periprocedural complications; No complications or FA at 6-month follow-up

Table 1. *Cont.*

Author, Year	Patient Age, Sex	Type and Severity of Hemophilia (Baseline Factor Activity Level)	CHA$_2$DS$_2$-VASc Score/HAS-BLED Score	Comorbidities	Procedure/Device	Antithrombotic Treatment after the Procedure and on Long-Term	Coagulation Factor Replacement Treatment	Outcome
Güray et al., 2019 [95]	67 y, M	HA, (baseline FVIII activity level ~10% with rFVIII)	3/3	Hypertension, HF	Amplatzer Amulet	DAPT with aspirin + clopidogrel 1 mo; SAPT with aspirin 2 mo	Adequate FVIII prophylaxis	No complications at 1-year follow-up
Coppola et al., 2020 [96]	Elderly, M	HA, severe (<1%)	3/NR	Advanced arthropathy	Amplatzer Plug	SAPT with clopidogrel	FVIII level ≥80% during and 12 h after the procedure; >5% on SAPT	Clopidogrel stopped after 2 mo due to severe epistaxis and joint bleeds
	Elderly, M	HA, severe (<1%)	3/NR	Advanced arthropathy	Amplatzer Plug	SAPT with clopidogrel	FVIII level ≥80% during and 12 h after the procedure; >5% on SAPT	NR
	76 y, M	HA, severe (<1%)	3/3	Hypertension	Amplatzer Amulet	SAPT with clopidogrel 3 mo	FVIII level >60% before the procedure	Minor hemarthrosis and epistaxis while on SAPT No complications at 20-month follow-up
Toselli et al., 2020 [67]	73 y, M	HB, moderate	4/3	Cardiac bypass surgery, HF (LVEF 40%), hip replacement surgery	Amplatzer Amulet	DAPT 3 mo SAPT lifelong	FIX before the procedure	No complications at 12-month follow-up
	79 y, M	HA, severe (<1%)	5/6	TIA, recurrent spontaneous hemarthroses, chronic kidney insufficiency, HCV-related chronic liver disease, treated hepatocellular carcinoma	Amplatzer Amulet	DAPT 3 wk SAPT 3 mo	FVIII level 65% before the procedure	Postprocedural acute pericarditis and mild transitory acute renal injury
Santoro et al., 2021 [97]	69 y, M	HB, moderate (3.5%)	3/3	DES for ACS, hypertension, melena and severe anemia while on DAPT and epistaxis while on SAPT— the patient refused FIX prophylaxis, hyper-homocysteinemia, curative treatment of low-grade transitional cell carcinoma, surgery for basal cell carcinoma	Left atrial appendage closure and cardioversion	Apixaban 2.5 mg bd, 1 mo DAPT with aspirin + clopidogrel 3 mo SAPT with clopidogrel lifelong	Eftrenonacog alfa	No postprocedural complications; No complications at 18-month follow-up

Table 1. *Cont.*

Author, Year	Patient Age, Sex	Type and Severity of Hemophilia (Baseline Factor Activity Level)	CHA$_2$DS$_2$-VASc Score/HAS-BLED Score	Comorbidities	Procedure/Device	Antithrombotic Treatment after the Procedure and on Long-Term	Coagulation Factor Replacement Treatment	Outcome
Lim et al., 2021 [54]	79 y, M	HA, mild (9%)	7/5	TIA, hypertension, PCI, atrioventricular node ablation and cardiac resynchronization therapy pacemaker, atherosclerotic calcifications at the carotid bifurcation and bulbs, HF (LVEF = 38%)	Watchman	VKA 6 wk SAPT with aspirin lifelong	FVIII level 100% for the procedure and 30% on VKA	No complications at 15-month follow-up
Dognin et al., 2021 [98]	61 y, M	HA, severe	2/1	NR	Watchman	No antithrombotic therapy	FVIII replacement	NR
Kramer, et al., 2021 [52]	70 y, F	HA, mild (14%)	5/3	Obesity, hypertension, HF	Amplatzer Amulet	SAPT with aspirin 6 mo	FVIII level >100% for the procedure; FVIII replacement 1–4 days after the procedure	Periprocedural arterial puncture
	75 y, M	HA, mild (20%)	2/2	Hypertension	Amplatzer Amulet	SAPT with aspirin 6 mo		Minor access-site hematoma and bleeding
	76 y, M	HA, mild (21%)	3/3	Hypertension	Amplatzer Amulet	SAPT with aspirin 5 mo		Self-limiting pericardial effusion Aspirin stopped due to major GI bleeding
	65 y, M	HA, mild (38%)	2/2	Hypertension	Watchman	SAPT with aspirin 6 mo		No complication
	60 y, M	HA, moderate (4%)	1/1	HF	Watchman	SAPT with aspirin 6 mo		Minor access-site hematoma and bleeding
	74 y, M	HA, mild (30%)	3/2	Hypertension	Watchman	DAPT 3 mo SAPT with aspirin 9 mo		Minor access-site hematoma and bleeding
	78 y, M	HA, severe (<1%)	6/3	Stroke, recent AMI, DES implantation (2 mo previously)	Watchman	DAPT 1 mo SAPT with aspirin lifelong		Significant access-site bleeding Exitus 9 mo after LAAO due to staphylococcal sepsis and mitral valve endocarditis

HA = hemophilia A, HB = hemophilia B, VKA = vitamin K antagonist, SAPT = single antiplatelet therapy, DAPT = dual antiplatelet therapy, PVI = pulmonary vein isolation,
FVIII = coagulation factor VIII, FIX = coagulation factor IX, NYHA = New York Heart Association, TIA = transient ischemic attack, LVEF = left ventricular ejection fraction,
HCV = hepatitis C virus, DES = drug-eluting stent, ACS = acute coronary syndrome, PCI = percutaneous coronary intervention, AMI = acute myocardial infarction, HF = heart failure, LAAO = left atrial appendage occlusion, COPD = chronic obstructive pulmonary disease, CABG = coronary artery bypass grafting, ICA = internal carotid artery, GI= gastrointestinal; wk = week, mo = month, NR = not reported.

7. Percutaneous Left Atrial Appendage Occlusion

As more than 90% of emboli originate in the left atrial appendage (LAA) [99], percutaneous left atrial appendage occlusion (LAAO) arose from the need to provide protection from ischemic stroke and systemic embolism in AF patients for whom long-term anticoagulant treatment is contraindicated. Data provided by several studies confirmed that LAAO is a feasible, effective and safe alternative to oral anticoagulant treatment [100]. LAAO has comparable efficacy with VKA at a lower rate of major bleedings, particularly hemorrhagic stroke [101,102]. This strategy provides a 90% reduction in the rate of hemorrhagic stroke [103]. In high-risk AF patients, LAAO has comparable efficacy with NOAC in stroke prevention and a similar or better safety profile [104,105]. To prevent device-related thrombosis after implantation, antithrombotic treatment is required until the complete endothelialization of the device surface. In patients without contraindications to oral anticoagulation, the postprocedural antithrombotic therapy consists of 45 days of VKA/NOAC and aspirin, followed by six months of dual antiplatelet therapy (DAPT) with aspirin and clopidogrel and then lifelong single antiplatelet therapy (SAPT) with aspirin [100].

Patients who have had bleeding complications or who have contraindications to long-term oral anticoagulant treatment will receive DAPT with aspirin and clopidogrel for 1–6 months, followed by lifelong SAPT with aspirin [100]. In naïve patients, loading doses of aspirin and clopidogrel are indicated. Patients at extremely high risk of bleeding should receive SAPT for at least 2–4 weeks. If no antithrombotic drug can be administered, epicardial LAA occlusion or thoracoscopic LAA clipping should be considered instead of LAA occluder implantation [100]. Early data from an ongoing study suggest that reduced doses of rivaroxaban (10 mg od and 15 mg od) may be an alternative to postprocedural DAPT [106].

Watchman, Amplatzer Cardiac Plug, and Amulet devices are the most widely used (Table 1). The Amplatzer device family is less thrombogenic than the Watchman device [107], therefore, postprocedural anticoagulation is never necessary [12]. The patients receiving Amplatzer Amulet devices did not have a higher risk of device-related thrombosis while on antiplatelets, especially those on SAPT [108]. In patients at high risk of stroke and contraindication to oral anticoagulant treatment, the implantation of Watchman device followed by DAPT was also an effective strategy [109]. The first case of LAAO in a hemophiliac with a high risk of stroke due to AF was published ten years ago [53]. Since then, many successful implantations of Amplatzer Cardiac Plug and Amplatzer Amulet devices have been reported [53,67,95,96]. Patients had high ischemic and bleeding risks. Their CHA$_2$DS$_2$-VASc score ranged from three to six and the HAS-BLED score from three to six. The postprocedural antithrombotic treatment consisted of antiplatelet agents.

Successful use of the Watchman device was also reported in hemophiliacs with AF [54,98]. Postprocedural antithrombotic therapy varied widely, from anticoagulation to no treatment. In a 60-year-old HB patient with mild disease, AF catheter ablation and placement of the Watchman device were performed concomitantly in order to minimize the patient's exposure to the anticoagulant [94]. After the procedure, the patients received only one month of VKA treatment. Still, the outcome was favorable at a 6-month follow-up.

A combined therapy consisting of LAA closure and cardioversion was successfully performed in a 69-year-old patient with moderate HB while on apixaban 2.5 mg bid and prophylaxis with eftrenonacog alfa [97]. After the procedure, he received one-month apixaban, three months DAPT, and long-term clopidogrel with favorable outcomes.

The largest series of cases of percutaneous LAAO published to date included seven patients with HA [52]. Three patients received the Amplatzer Amulet device and the rest the Watchman device. Before the procedure, two patients were on NOAC, one on LMWH, and one on DAPT and LMWH. After the device implantation, only antiplatelets were recommended, mainly SAPT with aspirin. In two cases, short-term DAPT was used.

Procedural risks associated with LAAO are cardiac tamponade (2–4%), stroke (1–2%), and inguinal hematoma (1%). Device embolization is extremely rare (<1%), but may require emergency cardiac surgery [103,110]. Most complications in hemophiliacs were minor or significant bleedings at the site of venous puncture. In one case, periprocedural arterial puncture occurred and in two cases, pericardial effusion without tamponade was diagnosed [52,67].

Although periprocedural therapy varies between treating centers, prophylactic replacement of the coagulation factor is always necessary, as a very high factor level of $\geq 80\%$ must be achieved during the procedure [52–54]. This level of coagulation factor is the same as that recommended for major surgery [57]. The percutaneous LAAO procedure is minimally invasive and does not require such intense replacement therapy, but the high level is considered necessary due to the risk of intracardiac lesions and the possible conversion to thoracic surgery. Moreover, full heparinization is used during the procedure [52,54,94]. In the first three days after the procedure, a trough level of the coagulation factor of 80% was frequently used [52,53].

8. Long-Term Antithrombotic Treatment

Current guidelines do not support the use of antiplatelet therapy alone for stroke prevention in AF patients in the general population [12]. It was proven that aspirin alone has limited or no protective effect against ischemic stroke [111,112]. Adding clopidogrel to aspirin reduced the stroke rate compared to aspirin alone, but major bleedings significantly increased [113]. Moreover, DAPT with aspirin and clopidogrel did not offer better protection from stroke than VKA at a similar rate of major bleedings [114].

AF patients requiring antithrombotic treatment receive an anticoagulant. VKA is more effective than aspirin in preventing ischemic stroke, with similar rates of major bleedings even in those older than 75 years [115]. Four NOACs—apixaban, dabigatran, edoxaban, and rivaroxaban—are currently indicated for stroke prevention in AF patients. A meta-analysis of their pivotal trials showed that NOACs reduced by 19% the risk of stroke or systemic embolism, with 51% in the risk of hemorrhagic stroke, and with 10% all-cause mortality comparative to VKA [116]. There was also a 14% reduction in the risk of major bleedings with NOACs, most coming from a 52% reduction in intracranial hemorrhages. Still, gastrointestinal bleedings increased by 25% with NOACs compared to warfarin. Of note, fatal bleedings are halved by using a NOAC versus VKA [117]. Due to the more favorable benefit–risk ratio, NOACs are preferred over VKAs in patients with AF eligible for a NOAC—without prosthetic mechanical heart valves or moderate-severe mitral stenosis [12].

In men with AF in the general population, the anticoagulant treatment should be considered if the CHA_2DS_2-VASc score is ≥ 1 [12]. Since there is no evidence that hemophiliacs are protected from cardioembolism by their underlying coagulation defect [57], the antithrombotic treatment should be considered in the presence of AF. Table 2 provides a practical algorithm for the antithrombotic treatment.

The most recent therapeutic algorithm was proposed by Schutgens et al. and is based on both the CHA_2DS_2-VASc score and the severity of hemophilia [8]. Among hemophiliacs with baseline FVIII/IX level $\geq 20\%$, the anticoagulant treatment is recommended only if the CHA_2DS_2-VASc score is ≥ 2. All oral anticoagulants are allowed, but the use of NOACs over VKA is encouraged because of the superior safety profile of NOACs to that of VKA, particularly regarding intracranial bleeding [116]. NOACs are preferred, especially in patients with HB because VKA reduces plasma FIX levels and increases the severity of hemophilia and consequently the risk of intracranial hemorrhage. While on VKA, these patients will need even more intensive replacement therapy, which is burdensome and may increase the frequency of inhibitor development. The high frequency of intravenous coagulation factor infusion and the related costs make long-term VKA use prohibitive [94].

The plasma FVIII/IX level considered safe for oral anticoagulation is $\geq 30\%$ [7,55]. Thus, only a few patients will benefit from anticoagulant treatment without requiring replacement therapy, namely the patients with very mild hemophilia and native factor activity level above this threshold. For hemophiliacs with more severe disease, this high threshold is difficult to maintain in the long-term because it requires very frequent administration of replacement products, often every 1–2 days [51]. Several studies have reported the safe use of anticoagulant treatment in hemophiliacs with FVIII/IX levels $\geq$20% [118,119] including after catheter ablation [78]. Therefore, it has been proposed to lower the threshold from 30% to 20% [8].

Table 2. Practical guide for long-term antithrombotic treatment.

Parameter	Recommendation
1. Patient characteristics	
The risk of stroke and systemic embolism	The CHA$_2$DS$_2$-VASc risk score assessment If 0, long-term ACO treatment should not be offered. If $\geq$1, long-term ACO treatment is recommended.
The bleeding risk	The HAS-BLED risk score assessment An attempt will be made to reduce the risk by intervening on modifiable factors.
	Assessment of the severity of hemophilia Coagulation factor level Coagulation factor replacement therapy Mandatory levels: $\geq$20% while on ACO $\geq$5–10% while on DAPT, SAPT
	The presence of inhibitors is a contraindication for ACO
2. Therapeutic intervention	
Rate control	The ACO indication is based on CHA$_2$DS$_2$-VASc risk score
Ablation	2 mo ACO, then the ACO indication is based on CHA$_2$DS$_2$-VASc risk score
LAAO	DAPT 1–6 mo, then lifelong SAPT
3. The anticoagulant treatment	
Type	NOAC preferred over VKA in HA patients NOAC preferred in HB patients
Dose	Low dose NOAC Apixaban 2.5 mg bid Dabigatran 110 mg bid Edoxaban 30 mg od Rivaroxaban 10 mg od VKA for INR 2–3
4. Patient preferences	
The patient should be informed of the advantages and disadvantages of the proposed treatments.	

ACO = anticoagulant, DAPT = dual antiplatelet therapy, SAPT = single antiplatelet therapy, mo = month, LAAO = left atrial appendage occlusion, NOAC = non-vitamin K oral anticoagulant, VKA = vitamin K antagonist, HA = hemophilia A, HB = hemophilia B.

Dabigatran is a direct thrombin inhibitor. In its pivotal trial, the dose of 150 mg bid reduced the risk of stroke by 36% compared to VKA at a similar rate of major bleedings. The dose of 110 mg bid was associated with a similar risk of stroke and a lower rate of major bleedings [120]. Of note, both regimes were associated with lower rates of hemorrhagic stroke than VKA, but the 150 mg bid dose increased the incidence of gastrointestinal bleeding by 50%. Dabigatran was successfully administered in hemophiliacs with AF eligible for anticoagulant treatment [8,121]. The 110 mg bid regimen was used due to its similar efficacy to VKA at a lower risk of bleeding.

Apixaban is a direct inhibitor of FXa, more efficient and safe than VKA in preventing stroke or systemic embolism in patients with AF eligible for a NOAC [122]. A recent real-world data analysis evaluated the efficacy and safety of apixaban according to its dose [123]. The rates of stroke/systemic embolism were reduced by 30% with the 5 mg bid regimen and by 37% with the 2.5 mg bid regimen. The rates of major bleeding decreased by 41% for each of the two regimes. To date, the successful use of low-dose apixaban has been reported in hemophiliacs with AF [52,97].

Rivaroxaban is a direct inhibitor of FXa, non-inferior to VKA in preventing stroke and systemic embolism in patients with AF eligible for a NOAC and with similar overall safety [124]. In the pivotal trial, the intracranial bleedings were less frequent with rivaroxaban than VKA, but the number of gastrointestinal bleedings increased. A recent real-world dose-based data analysis of rivaroxaban showed that the 20 mg od regimen reduced the risk of stroke and systemic embolism and was associated with fewer major bleedings compared to VKA. The 15 mg od regimen was as effective and safer than VKA [125]. In Asians with AF, the efficacy and safety of the 10 mg dose were also assessed [126]. Data from studies conducted on the Asian population showed that rivaroxaban was associated with a lower risk of stroke or systemic embolism than warfarin, regardless of dose [127]. In patients with mild hemophilia, the 10 mg od regimen was used [52].

Edoxaban is a direct inhibitor of FXa non-inferior to VKA in preventing stroke or systemic embolism in patients with AF eligible for a NOAC, with significantly lower rates of major bleeding than VKA for both 60 mg od and 30 mg od regimen [128]. The 60 mg od regimen was slightly more efficient than VKA. The rates of life-threatening bleeding, intracranial bleeding, and major bleeding plus clinically relevant non-major bleeding were favorable to edoxaban for both regimes, except for gastrointestinal bleedings that were higher with 60 mg od edoxaban than with VKA and lower with 30 mg od edoxaban than with VKA.

Thus, the preference for NOACs in hemophiliacs with AF is justified as they provide protection against ischemic stroke at least as well as VKA, but with a lower hemorrhagic risk including reduction in intracranial bleeding.

Hemophiliacs with baseline FVIII/IX level 1–19% and patients with the severe disease under FVIII/IX prophylaxis are candidates for catheter ablation or antiplatelet therapy with aspirin only if the CHA_2DS_2-VASc score ≥ 4 [8]. As VKA doubles the risk of intracranial hemorrhage compared to aspirin [129], it is considered unsuitable in hemophiliacs with basal FVIII/IX level <20% [8]. The plasma FVIII/IX level considered safe for SAPT is ≥ 5–10% [1,7,55].

Of note, apixaban is the only NOAC that has been compared in a large randomized clinical trial with aspirin in AF patients considered unsuitable to receive a VKA [130]. The rate of stroke/systemic embolism was more than halved, while the risk of major bleeding or intracranial hemorrhage did not significantly increase. Thus, the administration of apixaban could be a solution for those hemophiliacs receiving aspirin instead of VKA because their risk of bleeding is considered too high or intensive prophylactic replacement therapy is not feasible [75].

The oral anticoagulant treatment is not recommended in hemophiliacs with severe disease without prophylaxis because the thrombin generation is comparable to that in patients with a therapeutic INR [131]. In hemophilia patients with inhibitors, the antithrombotic therapy is generally contraindicated due to the less predictable hemostatic response to bypassing agents [66].

9. Discussion

Modern replacement therapies of the deficient coagulation factor prolong the lifespan of hemophiliacs, enabling the occurrence of diseases associated with old age. Atherosclerotic cardiovascular disease challenges most physicians' ability to provide adequate care because it often requires antithrombotic therapy. Hemophilia is a rare disease, so there are no conditions for conducting randomized clinical trials. Therefore, the therapeutic approach to cardiovascular diseases in hemophiliacs including AF is based on case reports, observational studies, and the consensus of experts [1,7–9,55], which is unlikely to change in the near future.

The management of patients with AF relies on three pillars: avoiding stroke, better symptom management, and cardiovascular and comorbidity optimization [12]. The CHA$_2$DS$_2$-VASc and HAS-BLED scores are recommended for the assessment of the risk of stroke and bleeding in the general population, respectively. Stroke prevention has evolved considerably in the last decade. Anticoagulant therapy with VKA or better with NOAC is the mainstream treatment. Patients with contraindication to oral anticoagulants or at risk of bleeding under anticoagulant therapy perceived as too high to be acceptable can now be candidates for left atrial appendage occlusion/exclusion [12]. Rhythm control to reduce the AF related symptoms and to improve the quality of life can be implemented through drugs or more recently, cardiac ablation. Pulmonary vein isolation is usually performed, but surgical or thoracoscopic ablation is also available in selected cases.

Although used to guide treatment in hemophiliacs with AF [7,8], the CHA$_2$DS$_2$-VASc score has not been prospectively validated in this population. Still, high values of this score have been found in hemophiliacs [52–54,67], suggesting that the risk of stroke and systemic embolism may be higher than anticipated [13]. The exact risk of stroke/systemic embolism in hemophiliacs with AF is still unknown, but it is certain that it is present even in patients with severe hemophilia. The HAS-BLED score was rarely used in hemophiliacs to estimate the bleeding risk because it was believed that the coagulation factor deficiency itself carries a high risk of bleeding. Moreover, it has been hypothesized that this score underestimates the bleeding risk in these patients [50,68]. However, recent evidence showed that a HAS-BLED score $\geq$3 was associated with increased bleeding risk even in hemophiliacs [51].

The antithrombotic treatment increases the risk of bleeding in all hemophiliacs, regardless of the severity of the disease. It was found that during the anticoagulant treatment, the bleeding risk of hemophiliacs with the mild disease increased eight times compared to the controls [51]. In hemophiliacs eligible for long-term oral anticoagulants, a preference for NOAC instead of VKA is highlighted, justified by their better efficacy and safety profile. Moreover, low doses of NOACs prevent stroke equal to or more than warfarin, with less major bleedings [120,122–128]. According to the available data, apixaban and 10 mg rivaroxaban seem to be the most suitable to recommend to hemophiliacs presenting AF [75,127,132]. Reports of successful use of NOACs in hemophiliacs with AF have already begun to appear [8,52,97].

HB patients benefit greatly from these recommendations. While a large fraction of factor VIII in blood originates from liver sinusoidal endothelial cells [133], FIX is a vitamin K-dependent coagulation factor produced in the liver. The treatment with VKA will further reduce plasma FIX levels, increasing the severity of hemophilia and the need for substitution therapy. Moreover, FIX replacement therapy may interfere with VKA and make INR value unreliable [55]. Since the coagulation factors II, VII, and X are also vitamin K-depending coagulation factors synthesized in the liver, VKA administration has the potential to further aggravate the imbalance of the hemostasis in hemophiliacs.

The experience gained so far suggests that the anticoagulant therapy is safe for hemophiliacs under appropriate replacement therapy, so working with a hematologist is vital for therapeutic success. In general, levels $\geq$80% and aiming at 100% are recommended intraprocedural and up to three days thereafter [52–54,78], $\geq$20% under anticoagulant treatment and DAPT [8,9,78], and $\geq$5–10% under SAPT [1,7,55].

The management of AF in patients with hemophilia can be quite complex, particularly for the reduction in the risk of stroke in the long-term and in the context of invasive procedures. Since the anticoagulant treatment is considered safe at FVIII/FIX level $\geq$20% [8], few hemophiliacs can receive it without coagulation factor replacement therapy. Many will require frequent administrations, possibly every 1–2 days, which is burdensome in the long run. It should be noted that this type of replacement therapy is also very expensive. Avoiding the need for long-term anticoagulant therapy should be a priority, therefore, the optimal therapeutic solution must be sought with great care. Cardioversion helps to control AF-related symptoms, improves left ventricular function, and quality of life, but long-term anticoagulant therapy is often needed [12].

PVI with short-term replacement therapy during periprocedural treatment with VKA or NOAC seems to be a good option [8]. NOACs appear safer than bridging VKA with LMWH for PVI [78]. Catheter ablation proved to be effective for rhythm control and safe to perform in hemophiliacs [77,78]. Although the mandatory duration of the anticoagulant treatment is two months, it is recommended that the treatment be extended in patients with high ischemic risk [12]. Still, it must also be considered that by continuing the oral anticoagulant treatment beyond three months from the procedure, the thrombotic events will slightly decrease while the major bleedings will increase [134].

Percutaneous LAAO is another possibility for these patients with a high bleeding risk [135]. It attracts increasing interest due to its favorable antithrombotic therapy profile, namely, the possibility of using short-time DAPT or anticoagulant, then lifetime SAPT. While DAPT requires the same level of coagulation factor as the anticoagulant treatment [9], SAPT may be safely recommended in hemophiliacs with factor level $\geq$5–10% [1,7,55]. This is a feasible approach in hemophiliacs. Thus, advances in the treatment of hemophilia allow for the implementation of the modern therapeutic solutions of AF in hemophilic patients.

This study has several limitations. First, there are little data on the management of AF in hemophiliacs, especially on percutaneous cardiac ablation and LAA occlusion techniques, therefore the conclusions cannot be generalized. Second, studies that have reported the use of antithrombotic therapy in hemophiliacs have often included, in addition to patients with AF, those with acute coronary syndromes or elective coronary procedures (coronary artery bypass grafting and percutaneous coronary intervention with stent implantation), which made it difficult to extract the necessary information. Moreover, data on the management of AF in hemophiliacs have sometimes been reported along with those from patients with other hereditary bleeding disorders.

10. Conclusions

Patients with hemophilia are often perceived as having only an increased risk of bleeding. However, in the presence of AF and ischemic risk factors, they also have an increased risk of stroke and systemic embolism. The choice of appropriate therapy is greatly hampered by the lack of guidelines based on strong evidence. Although the physicians treat hemophiliacs with AF on a case-by-case basis, it should be highlighted that the treatment offered in the general population can be implemented in hemophiliacs if appropriate replacement therapy can be provided. As far as we know, our work provides the most complete, structured, and updated analysis of the current therapeutic approach of AF in hemophiliacs.

Author Contributions: Conceptualization, M.C.B., O.V.B., M.F., and C.R.; Methodology, D.P., I.B., and O.N.B.-F.; Writing—original draft preparation, M.C.B., L.I.B., and M.F.; Writing—review and editing, M.C.B., O.V.B., and L.I.B.; Supervision, M.C., I.-I.C., and C.R. All authors have read and agreed to the published version of the manuscript.

Funding: This research received no external funding.

Institutional Review Board Statement: Not applicable.

Informed Consent Statement: Not applicable.

Conflicts of Interest: The authors declare no conflict of interest.

References

1. Mannucci, P.M. Management of antithrombotic therapy for acute coronary syndromes and atrial fibrillation in patients with hemophilia. *Expert Opin. Pharmacother.* **2012**, *13*, 505–510. [CrossRef] [PubMed]
2. Darby, S.C.; Kan, S.W.; Spooner, R.J.; Giangrande, P.L.; Hill, F.G.; Hay, C.R.; Lee, C.A.; Ludlam, C.A.; Williams, M. Mortality rates, life expectancy, and causes of death in people with hemophilia A or B in the United Kingdom who were not infected with HIV. *Blood* **2007**, *110*, 815–825. [CrossRef] [PubMed]
3. Plug, I.; Van Der Bom, J.G.; Peters, M.; Mauser-Bunschoten, E.P.; De Goede-Bolder, A.; Heijnen, L.; Smit, C.; Willemse, J.; Rosendaal, F.R. Mortality and causes of death in patients with hemophilia, 1992-2001: A prospective cohort study. *J. Thromb. Haemost.* **2006**, *4*, 510–516. [CrossRef] [PubMed]
4. Mannucci, P.M.; Mauser-Bunschoten, E.P. Cardiovascular disease in haemophilia patients: A contemporary issue. *Haemophilia* **2010**, *16* (Suppl. S3), 58–66. [CrossRef]
5. Philipp, C. The aging patient with hemophilia: Complications, comorbidities, and management issues. *Hematol. Am. Soc. Hematol. Educ. Program* **2010**, *2010*, 191–196. [CrossRef]
6. Mannucci, P.M. Hemophilia therapy: The future has begun. *Haematologica* **2020**, *105*, 545–553. [CrossRef]
7. Ferraris, V.A.; Boral, L.I.; Cohen, A.J.; Smyth, S.S.; White, G.C., 2nd. Consensus review of the treatment of cardiovascular disease in people with hemophilia A and B. *Cardiol. Rev.* **2015**, *23*, 53–68. [CrossRef]
8. Schutgens, R.E.; van der Heijden, J.F.; Mauser-Bunschoten, E.P.; Mannucci, P.M. New concepts for anticoagulant therapy in persons with hemophilia. *Blood* **2016**, *128*, 2471–2474. [CrossRef]
9. Schutgens, R.E.G.; Voskuil, M.; Mauser-Bunschoten, E.P. Management of cardiovascular disease in aging persons with haemophilia. *Hamostaseologie* **2017**, *37*, 196–201. [CrossRef]
10. Benjamin, E.J.; Muntner, P.; Alonso, A.; Bittencourt, M.S.; Callaway, C.W.; Carson, A.P.; Chamberlain, A.M.; Chang, A.R.; Cheng, S.; Das, S.R.; et al. Heart Disease and Stroke Statistics-2019 Update: A Report From the American Heart Association. *Circulation* **2019**, *139*, e56–e528. [CrossRef]
11. Krijthe, B.P.; Kunst, A.; Benjamin, E.J.; Lip, G.Y.; Franco, O.H.; Hofman, A.; Witteman, J.C.; Stricker, B.H.; Heeringa, J. Projections on the number of individuals with atrial fibrillation in the European Union, from 2000 to 2060. *Eur. Heart J.* **2013**, *34*, 2746–2751. [CrossRef] [PubMed]
12. Hindricks, G.; Potpara, T.; Dagres, N.; Arbelo, E.; Bax, J.J.; Blomstrom-Lundqvist, C.; Boriani, G.; Castella, M.; Dan, G.A.; Dilaveris, P.E.; et al. 2020 ESC Guidelines for the diagnosis and management of atrial fibrillation developed in collaboration with the European Association for Cardio-Thoracic Surgery (EACTS): The Task Force for the diagnosis and management of atrial fibrillation of the European Society of Cardiology (ESC) Developed with the special contribution of the European Heart Rhythm Association (EHRA) of the ESC. *Eur. Heart J.* **2021**, *42*, 373–498. [CrossRef] [PubMed]
13. Schutgens, R.E.; Klamroth, R.; Pabinger, I.; Malerba, M.; Dolan, G.; ADVANCE Working Group. Atrial fibrillation in patients with haemophilia: A cross-sectional evaluation in Europe. *Haemophilia* **2014**, *20*, 682–686. [CrossRef] [PubMed]
14. Pocoski, J.; Rule, B.; Ogbonnaya, A.; Lamerato, L.; Eaddy, M.; Lunacsek, O.; Humphries, T.J. Cardiovascular comorbidities in a United States patient population with hemophilia A: A comprehensive chart review. *Blood* **2016**, *128*, 4966. [CrossRef]
15. Staerk, L.; Sherer, J.A.; Ko, D.; Benjamin, E.J.; Helm, R.H. Atrial Fibrillation: Epidemiology, Pathophysiology, and Clinical Outcomes. *Circ. Res.* **2017**, *120*, 1501–1517. [CrossRef]
16. Schnabel, R.B.; Yin, X.; Gona, P.; Larson, M.G.; Beiser, A.S.; McManus, D.D.; Newton-Cheh, C.; Lubitz, S.A.; Magnani, J.W.; Ellinor, P.T.; et al. 50 year trends in atrial fibrillation prevalence, incidence, risk factors, and mortality in the Framingham Heart Study: A cohort study. *Lancet* **2015**, *386*, 154–162. [CrossRef]
17. Lip, G.Y.H.; Coca, A.; Kahan, T.; Boriani, G.; Manolis, A.S.; Olsen, M.H.; Oto, A.; Potpara, T.S.; Steffel, J.; Marin, F.; et al. Hypertension and cardiac arrhythmias: A consensus document from the European Heart Rhythm Association (EHRA) and ESC Council on Hypertension, endorsed by the Heart Rhythm Society (HRS), Asia-Pacific Heart Rhythm Society (APHRS) and Sociedad Latinoamericana de Estimulacion Cardiaca y Electrofisiologia (SOLEACE). *EP Eur.* **2017**, *19*, 891–911. [CrossRef]
18. Liang, F.; Wang, Y. Coronary heart disease and atrial fibrillation: A vicious cycle. *Am. J. Physiol. Heart Circ. Physiol.* **2021**, *320*, H1–H12. [CrossRef]
19. Thomas, K.L.; Jackson, L.R., 2nd; Shrader, P.; Ansell, J.; Fonarow, G.C.; Gersh, B.; Kowey, P.R.; Mahaffey, K.W.; Singer, D.E.; Thomas, L.; et al. Prevalence, Characteristics, and Outcomes of Valvular Heart Disease in Patients with Atrial Fibrillation: Insights From the ORBIT-AF (Outcomes Registry for Better Informed Treatment for Atrial Fibrillation). *J. Am. Heart Assoc.* **2017**, *6*, e006475. [CrossRef]
20. Carlisle, M.A.; Fudim, M.; DeVore, A.D.; Piccini, J.P. Heart Failure and Atrial Fibrillation, Like Fire and Fury. *JACC Heart Fail.* **2019**, *7*, 447–456. [CrossRef]
21. Cadby, G.; McArdle, N.; Briffa, T.; Hillman, D.R.; Simpson, L.; Knuiman, M.; Hung, J. Severity of OSA is an independent predictor of incident atrial fibrillation hospitalization in a large sleep-clinic cohort. *Chest* **2015**, *148*, 945–952. [CrossRef] [PubMed]
22. Boriani, G.; Savelieva, I.; Dan, G.A.; Deharo, J.C.; Ferro, C.; Israel, C.W.; Lane, D.A.; La Manna, G.; Morton, J.; Mitjans, A.M.; et al. Chronic kidney disease in patients with cardiac rhythm disturbances or implantable electrical devices: Clinical significance and implications for decision making-a position paper of the European Heart Rhythm Association endorsed by the Heart Rhythm Society and the Asia Pacific Heart Rhythm Society. *EP Eur.* **2015**, *17*, 1169–1196. [CrossRef]
23. Aune, D.; Schlesinger, S.; Norat, T.; Riboli, E. Tobacco smoking and the risk of atrial fibrillation: A systematic review and meta-analysis of prospective studies. *Eur. J. Prev. Cardiol.* **2018**, *25*, 1437–1451. [CrossRef]

24. Biere-Rafi, S.; Tuinenburg, A.; Haak, B.W.; Peters, M.; Huijgen, R.; De Groot, E.; Verhamme, P.; Peerlinck, K.; Visseren, F.L.; Kruip, M.J.; et al. Factor VIII deficiency does not protect against atherosclerosis. *J. Thromb. Haemost.* **2012**, *10*, 30–37. [CrossRef] [PubMed]

25. Foley, C.J.; Nichols, L.; Jeong, K.; Moore, C.G.; Ragni, M.V. Coronary atherosclerosis and cardiovascular mortality in hemophilia. *J. Thromb. Haemost.* **2010**, *8*, 208–211. [CrossRef]

26. Tuinenburg, A.; Rutten, A.; Kavousi, M.; Leebeek, F.W.; Ypma, P.F.; Laros-van Gorkom, B.A.; Nijziel, M.R.; Kamphuisen, P.W.; Mauser-Bunschoten, E.P.; Roosendaal, G.; et al. Coronary artery calcification in hemophilia A: No evidence for a protective effect of factor VIII deficiency on atherosclerosis. *Arter. Thromb. Vasc. Biol.* **2012**, *32*, 799–804. [CrossRef] [PubMed]

27. Zwiers, M.; Lefrandt, J.D.; Mulder, D.J.; Smit, A.J.; Gans, R.O.; Vliegenthart, R.; Brands-Nijenhuis, A.V.; Kluin-Nelemans, J.C.; Meijer, K. Coronary artery calcification score and carotid intima-media thickness in patients with hemophilia. *J. Thromb. Haemost.* **2012**, *10*, 23–29. [CrossRef] [PubMed]

28. Holme, P.A.; Combescure, C.; Tait, R.C.; Berntorp, E.; Rauchensteiner, S.; de Moerloose, P.; Group, A.W. Hypertension, haematuria and renal functioning in haemophilia—A cross-sectional study in Europe. *Haemophilia* **2016**, *22*, 248–255. [CrossRef]

29. van de Putte, D.E.F.; Fischer, K.; Makris, M.; Tait, R.C.; Chowdary, P.; Collins, P.W.; Meijer, K.; Roosendaal, G.; Schutgens, R.E.; Mauser-Bunschoten, E.P. Unfavourable cardiovascular disease risk profiles in a cohort of Dutch and British haemophilia patients. *Thromb. Haemost.* **2013**, *109*, 16–23. [CrossRef]

30. Wang, J.D.; Chan, W.C.; Fu, Y.C.; Tong, K.M.; Chang, S.T.; Hwang, W.L.; Lin, C.H.; Tsan, Y.T. Prevalence and risk factors of atherothrombotic events among 1054 hemophilia patients: A population-based analysis. *Thromb. Res.* **2015**, *135*, 502–507. [CrossRef]

31. Pocoski, J.; Ma, A.; Kessler, C.M.; Boklage, S.; Humphries, T.J. Cardiovascular comorbidities are increased in U.S. patients with haemophilia A: A retrospective database analysis. *Haemophilia* **2014**, *20*, 472–478. [CrossRef] [PubMed]

32. Sood, S.L.; Cheng, D.; Ragni, M.; Kessler, C.M.; Quon, D.; Shapiro, A.D.; Key, N.S.; Manco-Johnson, M.J.; Cuker, A.; Kempton, C.; et al. A cross-sectional analysis of cardiovascular disease in the hemophilia population. *Blood Adv.* **2018**, *2*, 1325–1333. [CrossRef] [PubMed]

33. Camelo, R.M.; Caram-Deelder, C.; Duarte, B.P.; de Moura, M.C.B.; Costa, N.C.M.; Costa, I.M.; Roncal, C.G.P.; Vanderlei, A.M.; Guimaraes, T.M.R.; Gouw, S.; et al. Cardiovascular risk factors among adult patients with haemophilia. *Int. J. Hematol.* **2021**, *113*, 884–892. [CrossRef] [PubMed]

34. Humphries, T.J.; Ma, A.; Kessler, C.M.; Kamalakar, R.; Pocoski, J. A second retrospective database analysis confirms prior findings of apparent increased cardiovascular comorbidities in hemophilia A in the United States. *Am. J. Hematol.* **2016**, *91*, E298–E299. [CrossRef]

35. Marchesini, E.; Oliovecchio, E.; Coppola, A.; Santagostino, E.; Radossi, P.; Castaman, G.; Valdre, L.; Santoro, C.; Tagliaferri, A.; Ettorre, C.; et al. Comorbidities in persons with haemophilia aged 60 years or more compared with age-matched people from the general population. *Haemophilia* **2018**, *24*, e6–e10. [CrossRef]

36. Magnussen, C.; Niiranen, T.J.; Ojeda, F.M.; Gianfagna, F.; Blankenberg, S.; Njolstad, I.; Vartiainen, E.; Sans, S.; Pasterkamp, G.; Hughes, M.; et al. Sex Differences and Similarities in Atrial Fibrillation Epidemiology, Risk Factors, and Mortality in Community Cohorts: Results from the BiomarCaRE Consortium (Biomarker for Cardiovascular Risk Assessment in Europe). *Circulation* **2017**, *136*, 1588–1597. [CrossRef]

37. van de Putte, D.E.F.; Fischer, K.; Makris, M.; Tait, R.C.; Chowdary, P.; Collins, P.W.; Meijer, K.; Roosendaal, G.; Schutgens, R.E.; Mauser-Bunschoten, E.P. History of non-fatal cardiovascular disease in a cohort of Dutch and British patients with haemophilia. *Eur. J. Haematol.* **2012**, *89*, 336–339. [CrossRef]

38. van de Putte, D.E.F.; Fischer, K.; Pulles, A.E.; Roosendaal, G.; Biesma, D.H.; Schutgens, R.E.; Mauser-Bunschoten, E.P. Non-fatal cardiovascular disease, malignancies, and other co-morbidity in adult haemophilia patients. *Thromb. Res.* **2012**, *130*, 157–162. [CrossRef]

39. Humphries, T.J.; Rule, B.; Ogbonnaya, A.; Eaddy, M.; Lunacsek, O.; Lamerato, L.; Pocoski, J. Cardiovascular comorbidities in a United States patient population with hemophilia A: A comprehensive chart review. *Adv. Med. Sci.* **2018**, *63*, 329–333. [CrossRef]

40. Violi, F.; Pastori, D.; Pignatelli, P. Mechanisms and Management of Thrombo-Embolism in Atrial Fibrillation. *J. Atr. Fibrillation* **2014**, *7*, 1112. [CrossRef]

41. Wolf, P.A.; Abbott, R.D.; Kannel, W.B. Atrial fibrillation as an independent risk factor for stroke: The Framingham Study. *Stroke* **1991**, *22*, 983–988. [CrossRef] [PubMed]

42. Furie, K.L.; Goldstein, L.B.; Albers, G.W.; Khatri, P.; Neyens, R.; Turakhia, M.P.; Turan, T.N.; Wood, K.A.; on behalf of the American Heart Association Stroke Council; Council on Quality of Care and Outcomes Research; et al. Oral antithrombotic agents for the prevention of stroke in nonvalvular atrial fibrillation: A science advisory for healthcare professionals from the American Heart Association/American Stroke Association. *Stroke* **2012**, *43*, 3442–3453. [CrossRef] [PubMed]

43. Andrew, N.E.; Thrift, A.G.; Cadilhac, D.A. The prevalence, impact and economic implications of atrial fibrillation in stroke: What progress has been made? *Neuroepidemiology* **2013**, *40*, 227–239. [CrossRef]

44. Camm, A.J.; Kirchhof, P.; Lip, G.Y.; Schotten, U.; Savelieva, I.; Ernst, S.; Van Gelder, I.C.; Al-Attar, N.; Hindricks, G.; Prendergast, B.; et al. Guidelines for the management of atrial fibrillation: The Task Force for the Management of Atrial Fibrillation of the European Society of Cardiology (ESC). *Eur. Heart J.* **2010**, *31*, 2369–2429. [CrossRef] [PubMed]

45. Fuster, V.; Ryden, L.E.; Cannom, D.S.; Crijns, H.J.; Curtis, A.B.; Ellenbogen, K.A.; Halperin, J.L.; Le Heuzey, J.Y.; Kay, G.N.; Lowe, J.E.; et al. ACC/AHA/ESC 2006 Guidelines for the Management of Patients with Atrial Fibrillation: A report of the American College of Cardiology/American Heart Association Task Force on Practice Guidelines and the European Society of Cardiology Committee for Practice Guidelines (Writing Committee to Revise the 2001 Guidelines for the Management of Patients

With Atrial Fibrillation): Developed in collaboration with the European Heart Rhythm Association and the Heart Rhythm Society. *Circulation* **2006**, *114*, e257–e354. [CrossRef]

46. You, J.J.; Singer, D.E.; Howard, P.A.; Lane, D.A.; Eckman, M.H.; Fang, M.C.; Hylek, E.M.; Schulman, S.; Go, A.S.; Hughes, M.; et al. Antithrombotic therapy for atrial fibrillation: Antithrombotic Therapy and Prevention of Thrombosis, 9th ed: American College of Chest Physicians Evidence-Based Clinical Practice Guidelines. *Chest* **2012**, *141*, e531S–e575S. [CrossRef]

47. Gage, B.F.; Waterman, A.D.; Shannon, W.; Boechler, M.; Rich, M.W.; Radford, M.J. Validation of clinical classification schemes for predicting stroke: Results from the National Registry of Atrial Fibrillation. *JAMA* **2001**, *285*, 2864–2870. [CrossRef]

48. Coppens, M.; Eikelboom, J.W.; Hart, R.G.; Yusuf, S.; Lip, G.Y.; Dorian, P.; Shestakovska, O.; Connolly, S.J. The CHA2DS2-VASc score identifies those patients with atrial fibrillation and a CHADS2 score of 1 who are unlikely to benefit from oral anticoagulant therapy. *Eur. Heart J.* **2013**, *34*, 170–176. [CrossRef]

49. Lip, G.Y.; Nieuwlaat, R.; Pisters, R.; Lane, D.A.; Crijns, H.J. Refining clinical risk stratification for predicting stroke and thromboembolism in atrial fibrillation using a novel risk factor-based approach: The euro heart survey on atrial fibrillation. *Chest* **2010**, *137*, 263–272. [CrossRef]

50. Schutgens, R.E.; Klamroth, R.; Pabinger, I.; Dolan, G.; Group, A.W. Management of atrial fibrillation in people with haemophilia-a consensus view by the ADVANCE Working Group. *Haemophilia* **2014**, *20*, e417–e420. [CrossRef]

51. Guillet, B.; Cayla, G.; Lebreton, A.; Trillot, N.; Wibaut, B.; Falaise, C.; Castet, S.; Gautier, P.; Claeyssens, S.; Schved, J.F. Long-Term Antithrombotic Treatments Prescribed for Cardiovascular Diseases in Patients with Hemophilia: Results from the French Registry. *Thromb. Haemost.* **2021**, *121*, 287–296. [CrossRef] [PubMed]

52. Kramer, A.D.; Korsholm, K.; Kristensen, A.; Poulsen, L.H.; Nielsen-Kudsk, J.E. Left atrial appendage occlusion in haemophilia patients with atrial fibrillation. *J. Interv. Card. Electrophysiol.* 2021, *ahead of print*. [CrossRef]

53. Cheung, V.T.; Hunter, R.J.; Ginks, M.R.; Schilling, R.J.; Earley, M.J.; Bowles, L. Management of thromboembolic risk in persons with haemophilia and atrial fibrillation: Is left atrial appendage occlusion the answer for those at high risk? *Haemophilia* **2013**, *19*, e84–e86. [CrossRef] [PubMed]

54. Lim, M.Y.; Abou-Ismail, M.Y. Left atrial appendage occlusion for management of atrial fibrillation in persons with hemophilia. *Thromb. Res.* **2021**, *206*, 9–13. [CrossRef] [PubMed]

55. Mannucci, P.M.; Schutgens, R.E.; Santagostino, E.; Mauser-Bunschoten, E.P. How I treat age-related morbidities in elderly persons with hemophilia. *Blood* **2009**, *114*, 5256–5263. [CrossRef] [PubMed]

56. Pisters, R.; Lane, D.A.; Nieuwlaat, R.; de Vos, C.B.; Crijns, H.J.; Lip, G.Y. A novel user-friendly score (HAS-BLED) to assess 1-year risk of major bleeding in patients with atrial fibrillation: The Euro Heart Survey. *Chest* **2010**, *138*, 1093–1100. [CrossRef]

57. Srivastava, A.; Santagostino, E.; Dougall, A.; Kitchen, S.; Sutherland, M.; Pipe, S.W.; Carcao, M.; Mahlangu, J.; Ragni, M.V.; Windyga, J.; et al. WFH Guidelines for the Management of Hemophilia, 3rd edition. *Haemophilia* **2020**, *26* (Suppl. S6), 1–158. [CrossRef] [PubMed]

58. de Tezanos Pinto, M.; Fernandez, J.; Perez Bianco, P.R. Update of 156 episodes of central nervous system bleeding in hemophiliacs. *Haemostasis* **1992**, *22*, 259–267. [CrossRef]

59. Ljung, R.C. Intracranial haemorrhage in haemophilia A and B. *Br. J. Haematol.* **2008**, *140*, 378–384. [CrossRef]

60. Stieltjes, N.; Calvez, T.; Demiguel, V.; Torchet, M.F.; Briquel, M.E.; Fressinaud, E.; Claeyssens, S.; Coatmelec, B.; Chambost, H.; French, I.C.H.S.G. Intracranial haemorrhages in French haemophilia patients (1991–2001): Clinical presentation, management and prognosis factors for death. *Haemophilia* **2005**, *11*, 452–458. [CrossRef]

61. Ghosh, K.; Nair, A.P.; Jijina, F.; Madkaikar, M.; Shetty, S.; Mohanty, D. Intracranial haemorrhage in severe haemophilia: Prevalence and outcome in a developing country. *Haemophilia* **2005**, *11*, 459–462. [CrossRef]

62. Traivaree, C.; Blanchette, V.; Armstrong, D.; Floros, G.; Stain, A.M.; Carcao, M.D. Intracranial bleeding in haemophilia beyond the neonatal period-the role of CT imaging in suspected intracranial bleeding. *Haemophilia* **2007**, *13*, 552–559. [CrossRef] [PubMed]

63. Stanton, R.; Demel, S.L.; Flaherty, M.L.; Antzoulatos, E.; Gilkerson, L.A.; Osborne, J.; Behymer, T.P.; Moomaw, C.J.; Sekar, P.; Langefeld, C.; et al. Risk of intracerebral haemorrhage from hypertension is greatest at an early age. *Eur. Stroke J.* **2021**, *6*, 28–35. [CrossRef] [PubMed]

64. Hay, C.R.; Palmer, B.; Chalmers, E.; Liesner, R.; Maclean, R.; Rangarajan, S.; Williams, M.; Collins, P.W.; on behalf of United Kingdom Haemophilia Centre Doctors' Organisation (UKHCDO). Incidence of factor VIII inhibitors throughout life in severe hemophilia A in the United Kingdom. *Blood* **2011**, *117*, 6367–6370. [CrossRef] [PubMed]

65. Lacroix-Desmazes, S.; Voorberg, J.; Lillicrap, D.; Scott, D.W.; Pratt, K.P. Tolerating Factor VIII: Recent Progress. *Front. Immunol.* **2019**, *10*, 2991. [CrossRef]

66. Martin, K.; Key, N.S. How I treat patients with inherited bleeding disorders who need anticoagulant therapy. *Blood* **2016**, *128*, 178–184. [CrossRef]

67. Toselli, M.; Bosi, D.; Benatti, G.; Solinas, E.; Cattabiani, M.A.; Vignali, L. Left atrial appendage closure: A balanced management of the thromboembolic risk in patients with hemophilia and atrial fibrillation. *J. Thromb. Thrombolysis* **2020**, *50*, 668–673. [CrossRef]

68. Cohen, O.C.; Bertelli, M.; Manmathan, G.; Little, C.; Riddell, A.; Pollard, D.; Aradom, E.; Mussara, M.; Harrington, C.; Kanagasabapathy, P.; et al. Challenges of antithrombotic therapy in the management of cardiovascular disease in patients with inherited bleeding disorders: A single-centre experience. *Haemophilia* **2021**, *27*, 425–433. [CrossRef]

69. de Vos, C.B.; Pisters, R.; Nieuwlaat, R.; Prins, M.H.; Tieleman, R.G.; Coelen, R.J.; van den Heijkant, A.C.; Allessie, M.A.; Crijns, H.J. Progression from paroxysmal to persistent atrial fibrillation clinical correlates and prognosis. *J. Am. Coll. Cardiol.* **2010**, *55*, 725–731. [CrossRef]

70. Ogawa, H.; An, Y.; Ikeda, S.; Aono, Y.; Doi, K.; Ishii, M.; Iguchi, M.; Masunaga, N.; Esato, M.; Tsuji, H.; et al. Progression from Paroxysmal to Sustained Atrial Fibrillation Is Associated With Increased Adverse Events. *Stroke* **2018**, *49*, 2301–2308. [CrossRef]
71. Dudink, E.; Erkuner, O.; Berg, J.; Nieuwlaat, R.; de Vos, C.B.; Weijs, B.; Capucci, A.; Camm, A.J.; Breithardt, G.; Le Heuzey, J.Y.; et al. The influence of progression of atrial fibrillation on quality of life: A report from the Euro Heart Survey. *EP Eur.* **2018**, *20*, 929–934. [CrossRef]
72. Bajpai, A.; Savelieva, I.; Camm, A.J. Treatment of atrial fibrillation. *Br. Med. Bull.* **2008**, *88*, 75–94. [CrossRef]
73. Shelton, R.J.; Clark, A.L.; Goode, K.; Rigby, A.S.; Houghton, T.; Kaye, G.C.; Cleland, J.G. A randomised, controlled study of rate versus rhythm control in patients with chronic atrial fibrillation and heart failure: (CAFE-II Study). *Heart* **2009**, *95*, 924–930. [CrossRef] [PubMed]
74. Murray, N.P.; Munoz, L.; Minzer, S.; Lopez, M.A. Management of Thrombosis Risk in a Carrier of Hemophilia A with Low Factor VIII Levels with Atrial Fibrillation: A Clinical Case and Literature Review. *Case Rep. Hematol.* **2018**, *2018*, 2615838. [CrossRef] [PubMed]
75. Aguilar, C. Antithrombotic therapy in a patient with mild haemophilia A and atrial fibrillation: Case report and brief review of the literature. *Blood Coagul. Fibrinolysis* **2015**, *26*, 346–349. [CrossRef] [PubMed]
76. Abdulla, K.H.; Tankut, S.S.; Doran, J.A.; Patel, P. Chemical cardioversion and aspirin prophylaxis: A novel management strategy for atrial fibrillation/flutter in a patient with hemophilia A. *J. Am. Coll. Cardiol.* **2020**, *75*, 2835. [CrossRef]
77. Lin, J.Y.; Igic, P.; Hoffmayer, K.S.; Field, M.E. Patients with hemophilia: Unique challenges for atrial fibrillation management. *HeartRhythm. Case Rep.* **2015**, *1*, 445–448. [CrossRef]
78. van der Valk, P.R.; Mauser-Bunschoten, E.P.; van der Heijden, J.F.; Schutgens, R.E.G. Catheter Ablation for Atrial Fibrillation in Patients with Hemophilia or von Willebrand Disease. *TH Open* **2019**, *3*, e335–e339. [CrossRef]
79. Sherman, D.G.; Kim, S.G.; Boop, B.S.; Corley, S.D.; Dimarco, J.P.; Hart, R.G.; Haywood, L.J.; Hoyte, K.; Kaufman, E.S.; Kim, M.H.; et al. Occurrence and characteristics of stroke events in the Atrial Fibrillation Follow-up Investigation of Sinus Rhythm Management (AFFIRM) study. *Arch. Intern. Med.* **2005**, *165*, 1185–1191. [CrossRef]
80. Steffel, J.; Verhamme, P.; Potpara, T.S.; Albaladejo, P.; Antz, M.; Desteghe, L.; Haeusler, K.G.; Oldgren, J.; Reinecke, H.; Roldan-Schilling, V.; et al. The 2018 European Heart Rhythm Association Practical Guide on the use of non-vitamin K antagonist oral anticoagulants in patients with atrial fibrillation. *Eur. Heart J.* **2018**, *39*, 1330–1393. [CrossRef]
81. Hill, K.; Sucha, E.; Rhodes, E.; Bota, S.; Hundemer, G.L.; Clark, E.G.; Canney, M.; Harel, Z.; Wang, Z.F.; Carrier, M.; et al. Amiodarone, verapamil, or diltiazem use with direct oral anticoagulants and the risk of hemorrhage in older adults. *CJC Open* **2021**, *4*, 315–323. [CrossRef]
82. Gandhi, S.K.; Reiffel, J.A.; Boiron, R.; Wieloch, M. Risk of Major Bleeding in Patients with Atrial Fibrillation Taking Dronedarone in Combination with a Direct Acting Oral Anticoagulant (From a U.S. Claims Database). *Am. J. Cardiol.* **2021**, *159*, 79–86. [CrossRef]
83. Hakalahti, A.; Biancari, F.; Nielsen, J.C.; Raatikainen, M.J. Radiofrequency ablation vs. antiarrhythmic drug therapy as first line treatment of symptomatic atrial fibrillation: Systematic review and meta-analysis. *EP Eur.* **2015**, *17*, 370–378. [CrossRef] [PubMed]
84. Calkins, H.; Hindricks, G.; Cappato, R.; Kim, Y.H.; Saad, E.B.; Aguinaga, L.; Akar, J.G.; Badhwar, V.; Brugada, J.; Camm, J.; et al. 2017 HRS/EHRA/ECAS/APHRS/SOLAECE expert consensus statement on catheter and surgical ablation of atrial fibrillation. *EP Eur.* **2018**, *20*, e1–e160. [CrossRef] [PubMed]
85. Oral, H.; Chugh, A.; Ozaydin, M.; Good, E.; Fortino, J.; Sankaran, S.; Reich, S.; Igic, P.; Elmouchi, D.; Tschopp, D.; et al. Risk of thromboembolic events after percutaneous left atrial radiofrequency ablation of atrial fibrillation. *Circulation* **2006**, *114*, 759–765. [CrossRef] [PubMed]
86. Hunter, R.J.; McCready, J.; Diab, I.; Page, S.P.; Finlay, M.; Richmond, L.; French, A.; Earley, M.J.; Sporton, S.; Jones, M.; et al. Maintenance of sinus rhythm with an ablation strategy in patients with atrial fibrillation is associated with a lower risk of stroke and death. *Heart* **2012**, *98*, 48–53. [CrossRef]
87. Themistoclakis, S.; Corrado, A.; Marchlinski, F.E.; Jais, P.; Zado, E.; Rossillo, A.; Di Biase, L.; Schweikert, R.A.; Saliba, W.I.; Horton, R.; et al. The risk of thromboembolism and need for oral anticoagulation after successful atrial fibrillation ablation. *J Am. Coll. Cardiol.* **2010**, *55*, 735–743. [CrossRef]
88. Nuhrich, J.M.; Kuck, K.H.; Andresen, D.; Steven, D.; Spitzer, S.G.; Hoffmann, E.; Schumacher, B.; Eckardt, L.; Brachmann, J.; Lewalter, T.; et al. Oral anticoagulation is frequently discontinued after ablation of paroxysmal atrial fibrillation despite previous stroke: Data from the German Ablation Registry. *Clin. Res. Cardiol.* **2015**, *104*, 463–470. [CrossRef]
89. Winkle, R.A.; Mead, R.H.; Engel, G.; Kong, M.H.; Patrawala, R.A. Atrial fibrillation ablation using open-irrigated tip radiofrequency: Experience with intraprocedural activated clotting times <=210 seconds. *Heart Rhythm.* **2014**, *11*, 963–968. [CrossRef]
90. Winkle, R.A.; Mead, R.H.; Engel, G.; Patrawala, R.A. Safety of lower activated clotting times during atrial fibrillation ablation using open irrigated tip catheters and a single transseptal puncture. *Am. J. Cardiol.* **2011**, *107*, 704–708. [CrossRef]
91. Kirchhof, P.; Benussi, S.; Kotecha, D.; Ahlsson, A.; Atar, D.; Casadei, B.; Castella, M.; Diener, H.C.; Heidbuchel, H.; Hendriks, J.; et al. 2016 ESC Guidelines for the management of atrial fibrillation developed in collaboration with EACTS. *Eur. Heart J.* **2016**, *37*, 2893–2962. [CrossRef]
92. McClure, G.R.; Belley-Cote, E.P.; Jaffer, I.H.; Dvirnik, N.; An, K.R.; Fortin, G.; Spence, J.; Healey, J.; Singal, R.K.; Whitlock, R.P. Surgical ablation of atrial fibrillation: A systematic review and meta-analysis of randomized controlled trials. *EP Eur.* **2018**, *20*, 1442–1450. [CrossRef] [PubMed]
93. Bogachev-Prokophiev, A.; Sharifulin, R.; Karadzha, A.; Larionova, N.; Shmyrev, V.; Kornilov, I.; Mamaev, A.; Afanasyev, A.; Pivkin, A. Minimally invasive mitral valve repair and ablation of concomitant atrial fibrillation in a patient with severe hemophilia A. *Clin Case Rep.* **2021**, *10*, e04174. [CrossRef]

94. Bhatti, Z.; Goldbarg, S. Combined left atrial appendage closure and ablation in a patient with hemophilia B, paroxysmal atrial fibrillation, and transient ischemic attack. *HeartRhythm Case Rep.* **2019**, *5*, 266–268. [CrossRef] [PubMed]
95. Guray, U.; Korkmaz, A.; Gursoy, H.T.; Elalmis, O.U. Percutaneous left atrial appendage closure in a patient with haemophilia and atrial fibrillation: A case report. *Eur. Heart. J. Case Rep.* **2019**, *3*, ytz124. [CrossRef]
96. Coppola, A.; Rivolta, G.F.; Quintavalle, G.; Matichecchia, A.; Riccardi, F.; Tagliaferri, A. Left atrial appendage closure in patients with atrial fibrillation and congenital bleeding disorders: A case-series. *Haemophilia* **2020**, *26*, 55. [CrossRef]
97. Santoro, R.C.; Falbo, M.; Ferraro, A. Apixaban and eftrenonacog alfa treatment in a patient with moderate hemophilia B and cardiovascular disease. *Hematol. Rep.* **2021**, *13*, 9169. [CrossRef]
98. Dognin, N.; Salaun, E.; Champagne, C.; Domain, G.; O'Hara, G.; Philippon, F.; Paradis, J.M.; Faroux, L.; Beaudoin, J.; O'Connor, K.; et al. Percutaneous left atrial appendage closure in patients with primary hemostasis disorders and atrial fibrillation. *J. Interv. Card. Electrophysiol.* 2021, *ahead of print*. [CrossRef]
99. Blackshear, J.L.; Odell, J.A. Appendage obliteration to reduce stroke in cardiac surgical patients with atrial fibrillation. *Ann. Thorac. Surg.* **1996**, *61*, 755–759. [CrossRef]
100. Glikson, M.; Wolff, R.; Hindricks, G.; Mandrola, J.; Camm, A.J.; Lip, G.Y.H.; Fauchier, L.; Betts, T.R.; Lewalter, T.; Saw, J.; et al. EHRA/EAPCI expert consensus statement on catheter-based left atrial appendage occlusion—An update. *EuroIntervention* **2020**, *15*, 1133–1180. [CrossRef]
101. Reddy, V.Y.; Doshi, S.K.; Kar, S.; Gibson, D.N.; Price, M.J.; Huber, K.; Horton, R.P.; Buchbinder, M.; Neuzil, P.; Gordon, N.T.; et al. 5-Year Outcomes After Left Atrial Appendage Closure: From the PREVAIL and PROTECT AF Trials. *J. Am. Coll. Cardiol.* **2017**, *70*, 2964–2975. [CrossRef]
102. Reddy, V.Y.; Sievert, H.; Halperin, J.; Doshi, S.K.; Buchbinder, M.; Neuzil, P.; Huber, K.; Whisenant, B.; Kar, S.; Swarup, V.; et al. Percutaneous left atrial appendage closure vs warfarin for atrial fibrillation: A randomized clinical trial. *JAMA* **2014**, *312*, 1988–1998. [CrossRef]
103. Holmes, D.R.; Reddy, V.Y.; Turi, Z.G.; Doshi, S.K.; Sievert, H.; Buchbinder, M.; Mullin, C.M.; Sick, P.; Investigators, P.A. Percutaneous closure of the left atrial appendage versus warfarin therapy for prevention of stroke in patients with atrial fibrillation: A randomised non-inferiority trial. *Lancet* **2009**, *374*, 534–542. [CrossRef]
104. Nielsen-Kudsk, J.E.; Korsholm, K.; Damgaard, D.; Valentin, J.B.; Diener, H.C.; Camm, A.J.; Johnsen, S.P. Clinical Outcomes Associated with Left Atrial Appendage Occlusion Versus Direct Oral Anticoagulation in Atrial Fibrillation. *JACC Cardiovasc. Interv.* **2021**, *14*, 69–78. [CrossRef]
105. Osmancik, P.; Herman, D.; Neuzil, P.; Hala, P.; Taborsky, M.; Kala, P.; Poloczek, M.; Stasek, J.; Haman, L.; Branny, M.; et al. Left Atrial Appendage Closure Versus Direct Oral Anticoagulants in High-Risk Patients with Atrial Fibrillation. *J. Am. Coll. Cardiol.* **2020**, *75*, 3122–3135. [CrossRef] [PubMed]
106. Duthoit, G.; Silvain, J.; Marijon, E.; Ducrocq, G.; Lepillier, A.; Frere, C.; Dimby, S.F.; Popovic, B.; Lellouche, N.; Martin-Toutain, I.; et al. Reduced Rivaroxaban Dose Versus Dual Antiplatelet Therapy After Left Atrial Appendage Closure: ADRIFT a Randomized Pilot Study. *Circ. Cardiovasc. Interv.* **2020**, *13*, e008481. [CrossRef] [PubMed]
107. Park, J.W.; Bethencourt, A.; Sievert, H.; Santoro, G.; Meier, B.; Walsh, K.; Lopez-Minguez, J.R.; Meerkin, D.; Valdes, M.; Ormerod, O.; et al. Left atrial appendage closure with Amplatzer cardiac plug in atrial fibrillation: Initial European experience. *Catheter. Cardiovasc. Interv.* **2011**, *77*, 700–706. [CrossRef] [PubMed]
108. Landmesser, U.; Tondo, C.; Camm, J.; Diener, H.C.; Paul, V.; Schmidt, B.; Settergren, M.; Teiger, E.; Nielsen-Kudsk, J.E.; Hildick-Smith, D. Left atrial appendage occlusion with the AMPLATZER Amulet device: One-year follow-up from the prospective global Amulet observational registry. *EuroIntervention* **2018**, *14*, e590–e597. [CrossRef] [PubMed]
109. Reddy, V.Y.; Mobius-Winkler, S.; Miller, M.A.; Neuzil, P.; Schuler, G.; Wiebe, J.; Sick, P.; Sievert, H. Left atrial appendage closure with the Watchman device in patients with a contraindication for oral anticoagulation: The ASAP study (ASA Plavix Feasibility Study with Watchman Left Atrial Appendage Closure Technology). *J. Am. Coll. Cardiol.* **2013**, *61*, 2551–2556. [CrossRef] [PubMed]
110. Ostermayer, S.H.; Reisman, M.; Kramer, P.H.; Matthews, R.V.; Gray, W.A.; Block, P.C.; Omran, H.; Bartorelli, A.L.; Della Bella, P.; Di Mario, C.; et al. Percutaneous left atrial appendage transcatheter occlusion (PLAATO system) to prevent stroke in high-risk patients with non-rheumatic atrial fibrillation: Results from the international multi-center feasibility trials. *J. Am. Coll. Cardiol.* **2005**, *46*, 9–14. [CrossRef]
111. Sjalander, S.; Sjalander, A.; Svensson, P.J.; Friberg, L. Atrial fibrillation patients do not benefit from acetylsalicylic acid. *EP Eur.* **2014**, *16*, 631–638. [CrossRef]
112. Lip, G.Y. The role of aspirin for stroke prevention in atrial fibrillation. *Nat. Rev. Cardiol.* **2011**, *8*, 602–606. [CrossRef]
113. ACTIVE Investigators; Connolly, S.J.; Pogue, J.; Hart, R.G.; Hohnloser, S.H.; Pfeffer, M.; Chrolavicius, S.; Yusuf, S. Effect of clopidogrel added to aspirin in patients with atrial fibrillation. *N. Engl. J. Med.* **2009**, *360*, 2066–2078. [CrossRef] [PubMed]
114. ACTIVE Writing Group of the ACTIVE Investigators; Connolly, S.; Pogue, J.; Hart, R.; Pfeffer, M.; Hohnloser, S.; Chrolavicius, S.; Pfeffer, M.; Hohnloser, S.; Yusuf, S. Clopidogrel plus aspirin versus oral anticoagulation for atrial fibrillation in the Atrial fibrillation Clopidogrel Trial with Irbesartan for prevention of Vascular Events (ACTIVE W): A randomised controlled trial. *Lancet* **2006**, *367*, 1903–1912. [CrossRef]
115. Mant, J.; Hobbs, F.D.; Fletcher, K.; Roalfe, A.; Fitzmaurice, D.; Lip, G.Y.; Murray, E.; Investigators, B. Midland Research Practices, N. Warfarin versus aspirin for stroke prevention in an elderly community population with atrial fibrillation (the Birmingham Atrial Fibrillation Treatment of the Aged Study, BAFTA): A randomised controlled trial. *Lancet* **2007**, *370*, 493–503. [CrossRef]

116. Ruff, C.T.; Giugliano, R.P.; Braunwald, E.; Hoffman, E.B.; Deenadayalu, N.; Ezekowitz, M.D.; Camm, A.J.; Weitz, J.I.; Lewis, B.S.; Parkhomenko, A.; et al. Comparison of the efficacy and safety of new oral anticoagulants with warfarin in patients with atrial fibrillation: A meta-analysis of randomised trials. *Lancet* **2014**, *383*, 955–962. [CrossRef]

117. Gomez-Outes, A.; Lagunar-Ruiz, J.; Terleira-Fernandez, A.I.; Calvo-Rojas, G.; Suarez-Gea, M.L.; Vargas-Castrillon, E. Causes of Death in Anticoagulated Patients with Atrial Fibrillation. *J. Am. Coll. Cardiol.* **2016**, *68*, 2508–2521. [CrossRef] [PubMed]

118. Tuinenburg, A.; Damen, S.A.; Ypma, P.F.; Mauser-Bunschoten, E.P.; Voskuil, M.; Schutgens, R.E. Cardiac catheterization and intervention in haemophilia patients: Prospective evaluation of the 2009 institutional guideline. *Haemophilia* **2013**, *19*, 370–377. [CrossRef] [PubMed]

119. Fogarty, P.F.; Mancuso, M.E.; Kasthuri, R.; Bidlingmaier, C.; Chitlur, M.; Gomez, K.; Holme, P.A.; James, P.; Kruse-Jarres, R.; Mahlangu, J.; et al. Presentation and management of acute coronary syndromes among adult persons with haemophilia: Results of an international, retrospective, 10-year survey. *Haemophilia* **2015**, *21*, 589–597. [CrossRef]

120. Connolly, S.J.; Ezekowitz, M.D.; Yusuf, S.; Eikelboom, J.; Oldgren, J.; Parekh, A.; Pogue, J.; Reilly, P.A.; Themeles, E.; Varrone, J.; et al. Dabigatran versus warfarin in patients with atrial fibrillation. *N. Engl. J. Med.* **2009**, *361*, 1139–1151. [CrossRef]

121. Serrano, R.; Dutra, R.; Arias, E.; Santos, A.; Antunes, D.M.J. Haemophilia and atrial fibrillation—A case report. *Haemophilia* **2021**, *27*, 82. [CrossRef]

122. Granger, C.B.; Alexander, J.H.; McMurray, J.J.; Lopes, R.D.; Hylek, E.M.; Hanna, M.; Al-Khalidi, H.R.; Ansell, J.; Atar, D.; Avezum, A.; et al. Apixaban versus warfarin in patients with atrial fibrillation. *N. Engl. J. Med.* **2011**, *365*, 981–992. [CrossRef]

123. Li, X.; Keshishian, A.; Hamilton, M.; Horblyuk, R.; Gupta, K.; Luo, X.; Mardekian, J.; Friend, K.; Nadkarni, A.; Pan, X.; et al. Apixaban 5 and 2.5 mg twice-daily versus warfarin for stroke prevention in nonvalvular atrial fibrillation patients: Comparative effectiveness and safety evaluated using a propensity-score-matched approach. *PLoS ONE* **2018**, *13*, e0191722. [CrossRef] [PubMed]

124. Patel, M.R.; Mahaffey, K.W.; Garg, J.; Pan, G.; Singer, D.E.; Hacke, W.; Breithardt, G.; Halperin, J.L.; Hankey, G.J.; Piccini, J.P.; et al. Rivaroxaban versus warfarin in nonvalvular atrial fibrillation. *N. Engl. J. Med.* **2011**, *365*, 883–891. [CrossRef]

125. Blin, P.; Fauchier, L.; Dureau-Pournin, C.; Sacher, F.; Dallongeville, J.; Bernard, M.A.; Lassalle, R.; Droz-Perroteau, C.; Moore, N. Effectiveness and Safety of Rivaroxaban 15 or 20 mg Versus Vitamin K Antagonists in Nonvalvular Atrial Fibrillation. *Stroke* **2019**, *50*, 2469–2476. [CrossRef] [PubMed]

126. Lin, Y.C.; Chien, S.C.; Hsieh, Y.C.; Shih, C.M.; Lin, F.Y.; Tsao, N.W.; Chen, C.W.; Kao, Y.T.; Chiang, K.H.; Chen, W.T.; et al. Effectiveness and Safety of Standard- and Low-Dose Rivaroxaban in Asians with Atrial Fibrillation. *J. Am. Coll. Cardiol.* **2018**, *72*, 477–485. [CrossRef] [PubMed]

127. Qian, J.; Yan, Y.D.; Yang, S.Y.; Zhang, C.; Li, W.Y.; Gu, Z.C. Benefits and Harms of Low-Dose Rivaroxaban in Asian Patients with Atrial Fibrillation: A Systematic Review and Meta-analysis of Real-World Studies. *Front. Pharmacol.* **2021**, *12*, 642907. [CrossRef]

128. Giugliano, R.P.; Ruff, C.T.; Braunwald, E.; Murphy, S.A.; Wiviott, S.D.; Halperin, J.L.; Waldo, A.L.; Ezekowitz, M.D.; Weitz, J.I.; Spinar, J.; et al. Edoxaban versus warfarin in patients with atrial fibrillation. *N. Engl. J. Med.* **2013**, *369*, 2093–2104. [CrossRef]

129. Hart, R.G.; Pearce, L.A.; Aguilar, M.I. Meta-analysis: Antithrombotic therapy to prevent stroke in patients who have nonvalvular atrial fibrillation. *Ann. Intern. Med.* **2007**, *146*, 857–867. [CrossRef]

130. Connolly, S.J.; Eikelboom, J.; Joyner, C.; Diener, H.C.; Hart, R.; Golitsyn, S.; Flaker, G.; Avezum, A.; Hohnloser, S.H.; Diaz, R.; et al. Apixaban in patients with atrial fibrillation. *N. Engl. J. Med.* **2011**, *364*, 806–817. [CrossRef]

131. de Koning, M.L.Y.; Fischer, K.; de Laat, B.; Huisman, A.; Ninivaggi, M.; Schutgens, R.E.G. Comparing thrombin generation in patients with hemophilia A and patients on vitamin K antagonists. *J. Thromb. Haemost.* **2017**, *15*, 868–875. [CrossRef]

132. Aguilar, C. Might apixaban be the optimal oral anticoagulant for haemophiliacs with atrial fibrillation? *Haemophilia* **2015**, *21*, e338–e340. [CrossRef]

133. Zanolini, D.; Merlin, S.; Feola, M.; Ranaldo, G.; Amoruso, A.; Gaidano, G.; Zaffaroni, M.; Ferrero, A.; Brunelleschi, S.; Valente, G.; et al. Extrahepatic sources of factor VIII potentially contribute to the coagulation cascade correcting the bleeding phenotype of mice with hemophilia A. *Haematologica* **2015**, *100*, 881–892. [CrossRef] [PubMed]

134. Karasoy, D.; Gislason, G.H.; Hansen, J.; Johannessen, A.; Kober, L.; Hvidtfeldt, M.; Ozcan, C.; Torp-Pedersen, C.; Hansen, M.L. Oral anticoagulation therapy after radiofrequency ablation of atrial fibrillation and the risk of thromboembolism and serious bleeding: Long-term follow-up in nationwide cohort of Denmark. *Eur. Heart J.* **2015**, *36*, 307–315. [CrossRef] [PubMed]

135. Holmes, D.R., Jr.; Kar, S.; Price, M.J.; Whisenant, B.; Sievert, H.; Doshi, S.K.; Huber, K.; Reddy, V.Y. Prospective randomized evaluation of the Watchman Left Atrial Appendage Closure device in patients with atrial fibrillation versus long-term warfarin therapy: The PREVAIL trial. *J. Am. Coll. Cardiol.* **2014**, *64*, 1–12. [CrossRef] [PubMed]

Journal of
*Personalized
Medicine*

Protocol

Supermarket/Hypermarket Opportunistic Screening for Atrial Fibrillation (SHOPS-AF): A Mixed Methods Feasibility Study Protocol

Ian D. Jones [1,2,*], Deirdre A. Lane [2,3,4], Robyn R. Lotto [1,2], David Oxborough [2,5], Lis Neubeck [6], Peter E. Penson [2,7], Gabriela Czanner [2,8], Andy Shaw [9], Emma Johnston Smith [1], Aimeris Santos [1,2], Emily E. McGinn [1,2], Aderonke Ajiboye [1,2], Nicola Town [1,2] and Gregory Y. H. Lip [2,3,4]

1. School of Nursing and Allied Health, Faculty of Health, Liverpool John Moores University, Liverpool L3 2AJ, UK; r.r.lotto@ljmu.ac.uk (R.R.L.); e.crawford@ljmu.ac.uk (E.J.S.); aimeris20@hotmail.com (A.S.); emcginn@hotmail.co.uk (E.E.M.); aderonke_ajiboye@yahoo.com (A.A.); n.j.town@ljmu.ac.uk (N.T.)
2. Liverpool Centre for Cardiovascular Science, Liverpool John Moores University, Liverpool L3 2AJ, UK; deirdre.lane@liverpool.ac.uk (D.A.L.); d.l.oxborough@ljmu.ac.uk (D.O.); p.penson@ljmu.ac.uk (P.E.P.); g.czanner@ljmu.ac.uk (G.C.); gregory.lip@liverpool.ac.uk (G.Y.H.L.)
3. Liverpool Centre for Cardiovascular Science, University of Liverpool and Liverpool Heart & Chest Hospital, Liverpool L7 8TX, UK
4. Department of Clinical Medicine, Aalborg University, 9000 Aalborg, Denmark
5. School of Sport and Exercise Science, Liverpool John Moores University, Liverpool L3 3AF, UK
6. School of Health and Social Care, Edinburgh Napier University, Edinburgh EH11 4DN, UK; l.neubeck@napier.ac.uk
7. School of Pharmacy and Biomolecular Sciences, Liverpool John Moores University, Liverpool L3 3AF, UK
8. School of Computer Science and Mathematics, Liverpool John Moores University, Liverpool L3 3AF, UK
9. School of Civil Engineering and Built Environment, Liverpool John Moores University, Liverpool L3 3AF, UK; a.shaw@ljmu.ac.uk
* Correspondence: i.d.jones@ljmu.ac.uk

Citation: Jones, I.D.; Lane, D.A.; Lotto, R.R.; Oxborough, D.; Neubeck, L.; Penson, P.E.; Czanner, G.; Shaw, A.; Johnston Smith, E.; Santos, A.; et al. Supermarket/Hypermarket Opportunistic Screening for Atrial Fibrillation (SHOPS-AF): A Mixed Methods Feasibility Study Protocol. *J. Pers. Med.* **2022**, *12*, 578. https://doi.org/10.3390/jpm12040578

Academic Editor: José Miguel Rivera-Caravaca

Received: 23 February 2022
Accepted: 28 March 2022
Published: 4 April 2022

Publisher's Note: MDPI stays neutral with regard to jurisdictional claims in published maps and institutional affiliations.

Abstract: Aims: Atrial fibrillation (AF) is the most common sustained cardiac arrhythmia and a key risk factor for ischaemic stroke. Following AF detection, treatment with oral anticoagulation can significantly lower mortality and morbidity rates associated with this risk. The availability of several hand-held devices which can detect AF may enable trained health professionals to adopt AF screening approaches which do not interfere with people's daily routines. This study aims to investigate the effectiveness of a hand-held device (the MyDiagnostick single-lead Electrocardiogram (ECG) sensor) in screening for AF when embedded into the handles of supermarket trolleys. Methods: A mixed methods two-phase approach will be taken. The quantitative first phase will involve the recruitment of 2000 participants from a convenience sample at four large supermarkets with pharmacies. Prospective participants will be asked to conduct their shopping using a trolley embedded with a MyDiagnostick sensor. If the device identifies a participant with AF, the in-store pharmacist will be dispatched to take a manual pulse measurement and a static control sensor reading and offer a cardiologist consultation referral. When the sensor does not detect AF, a researcher will confirm the reading with a manual pulse measurement. ECGs will be compiled, and the sensitivity, specificity and positive and negative predictive values will be determined. A qualitative second phase will consist of semi-structured interviews carried out with those pharmacists and store managers in-store during the running of the trial period. These will explore the perceptions of staff regarding the merits of embedding sensors in the handles of supermarket trolleys to detect AF. Conclusion: This feasibility study will inform a larger future definitive trial.

Keywords: atrial fibrillation detection; cardiac arrhythmia; stroke prevention; electrocardiogram; sensor devices; community screening

1. Introduction

Cardiovascular disease accounts for almost a quarter (24.7%) of all premature deaths in men and nearly a fifth (17%) in women in the U.K. [1]. Four percent of these deaths are stroke-related [1]. Indeed, while stroke mortality rates have reduced over time, the condition remains one of the major global causes of mortality and disability [2]. In the U.K., stroke is the third most common cause of years of life lost and of disability [3] and is estimated to cost the UK Health and Social Care system GBP 8.6 billion per annum [4]. This cost equates to GBP 22,429 in the first year, with ongoing five-year costs of GBP 46,039 per patient [2]. Eighty-seven percent of these admissions (126,640) are ischaemic stroke-related, and atrial fibrillation (AF) is responsible for around a quarter of such strokes, which are more likely to be severe and fatal compared to other causes of stroke [5,6]. Unlike many other cardiac conditions, recent UK registry-based data has noted that the prevalence of AF is increasing [1], which, if not treated effectively, will likely result in an increasing number of strokes in the longer term, with a major impact on NHS healthcare costs related to AF [7]. Moreover, Public Health England estimates that while around 123,000 people in England are living with AF, around 20% of these people are undiagnosed and are therefore untreated [8].

Whilst it is recognised that the risk of stroke in patients with AF is not homogeneous, all patients with AF should be risk-assessed using the CHA_2DS_2-VASc score [9,10], with most patients, except the very lowest risk (CHA_2DS_2-VASc score of 0 in men and 1 in women), receiving oral anticoagulation therapy. Anticoagulant therapy, using either vitamin K antagonists (VKAs) or non-vitamin K antagonist oral anticoagulants (NOACS), markedly reduces the risks of stroke and mortality in AF patients [11,12]. More recent focus has been placed on the detection and characterisation of AF patients, followed by a holistic approach to AF care, given the improved clinical outcomes with integrated care management [13,14].

The detection and diagnosis of AF requires ECG rhythm documentation showing the typical pattern of irregular RR intervals with an absence of discernible P waves [10]. By accepted convention, an episode lasting at least 30 s is considered diagnostic [10]. However, whilst an ECG provides documentary evidence, the presence of an irregular pulse arouses suspicion of AF and whilst not as specific or as sensitive as the ECG, checking an individual's pulse is much less costly and requires fewer resources. A pulse check is therefore an easy, low cost means of ruling out or raising suspicion of AF, as an approach to AF screening [15].

However, the regularity of a person's pulse and the normality of their ECG are usually only confirmed as a result of either opportunistic screening [10] or post-event identification. Silent or undetected AF is not uncommon [16,17], and although opportunistic screening is considered cost-effective in some populations [18], it may be seen as an inconvenience to those not currently experiencing symptoms and may be irrelevant to some socioeconomically deprived groups who do not engage with health professionals until a crisis occurs. Therefore, alternative models of screening that can be delivered by appropriately qualified personnel whilst not detracting people from their daily activities are required. Until recently, such a process has been impractical; however, with the advent of new technology, this may now be possible, as a number of validated hand-held devices are available that allow irregular pulses to be detected.

MyDiagnostick is a cylindrical shaped MDD Class IIa medical device (length 26 cm, diameter 2 cm) costing circa GBP 550 that records a single lead I ECG tracing when a participant's hands make contact with the metal electrodes (Figure 1). The device's algorithm-based internal software analyses the R-R interval of the ECG recording over a one-minute period whilst contact is maintained. The presence of AF is signified by a red light (Figure 2), whereas the absence of AF is signified by a green light (Figure 3). A single lead tracing can then be downloaded via a USB port onto a computer as a PDF document for further analysis. Whilst the MyDiagnostick algorithm is focussed solely on the presence or absence of AF, the PDF generated allows a trained professional to identify additional arrhythmias if present.

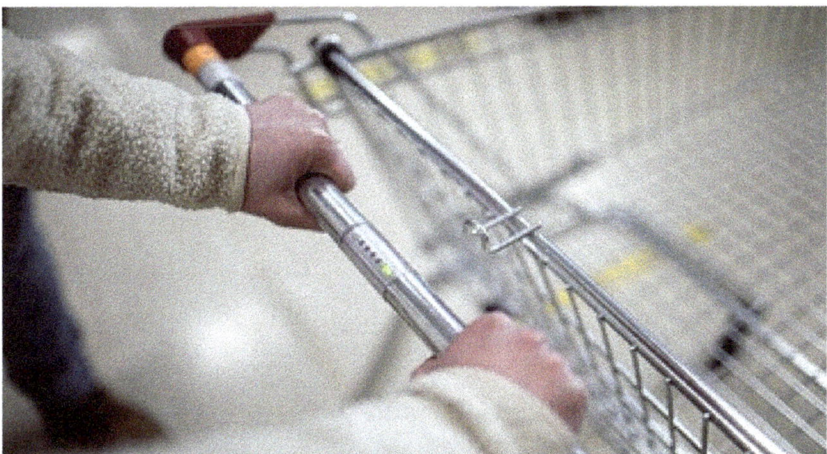

Figure 1. MyDiagnostick single lead ECG sensor embedded into the handles of supermarket trolleys to detect atrial fibrillation. A participant should place both hands around the device for 60 s. Depending on the detection of atrial fibrillation or not, the sensor will present with a red cross or a green tick, respectively.

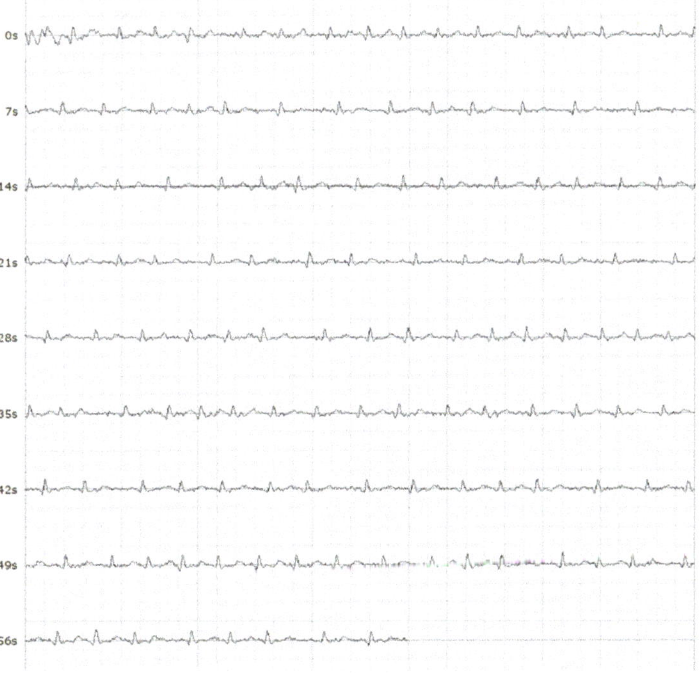

Figure 2. A single lead ECG tracing showing AF.

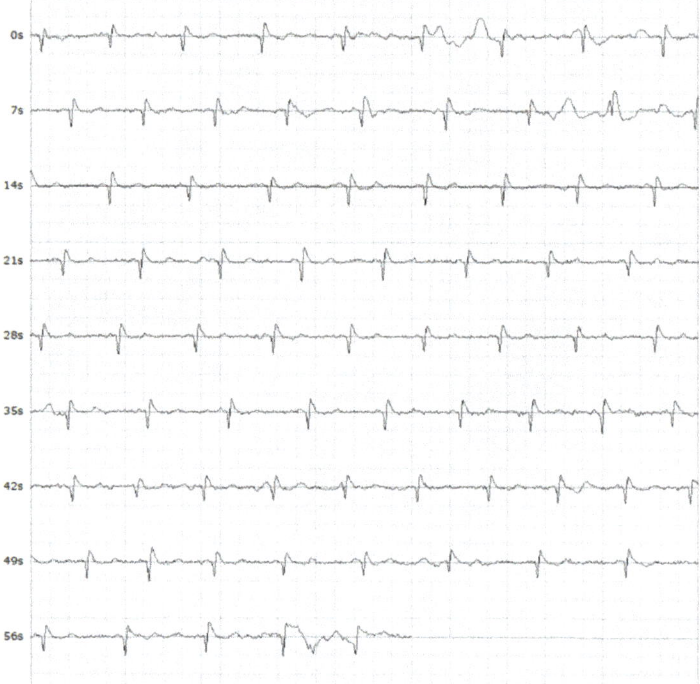

Figure 3. A single lead ECG tracing that was recorded as not AF.

We propose to embed a MyDiagnostick single lead ECG sensor into the handles of supermarket trolleys to assess the person's heart rhythm whilst they undertake their shopping (Figure 1). If an abnormal heart rhythm is detected, the store's pharmacist is notified and is dispatched to meet with the shopper. Community-based pharmacies have been identified as a potential location to undertake opportunistic screening and lifestyle interventions [19]. Previous studies have demonstrated that pharmacists are able to successfully screen and identify those at risk of cardiovascular disease [20,21] and is acceptable to patients [22], but again, these studies have relied on self-presentation.

2. Methods

We are proposing to locate a research team in four large supermarkets that contain pharmacies where shoppers will be asked to test the embedded technology, and store managers and pharmacists will provide their views on the merits of the screening approach. We will do this using a mixed methods two-phase approach.

2.1. Phase 1

2.1.1. Study Design

A cross-sectional observational study with a convenience sample will be used to address the research objectives.

2.1.2. Eligibility Criteria

Members of the public who are visiting four large supermarkets in the North West of England. We will recruit from four supermarkets for a period of two months each. To be eligible for inclusion, participants must be aged $\geq$18 years, able to grip a shopping trolley handle and provide written informed consent. Participants will be excluded if they have a physical tremor as this affects the ability of the device to record the heart rhythm due to movement artefact and previously participated in the study. Those with known AF will not be excluded from participation as the inclusion of this group of patients will aid the assessment of the sensitivity of the sensor and minimise selection bias.

2.1.3. Sample Size

Large supermarkets attract an average footfall of 25,000 people per week [23], providing a potential total population across four stores per week of 100,000 people. Recognising that many of these people are returning customers attending on a weekly basis, this study aims to recruit around 2% of the total population, resulting in a sample of 2000 participants. The prevalence of AF in the area of the North West of England from which participants are to be recruited is estimated to be 2.1% [24,25], which would result in 42/2000 positive results, including some who are known to experience AF. However, we recognise that those most at risk of developing AF are $\geq$ 65 years. Unfortunately, neither the British Retail Consortium nor YouGov provides age-related data for in-store supermarket shopping. One American study estimated that those over the age of 60 years make up 24% of supermarket consumers [26]. Recognising that the prevalence of AF is highest in this older age group, ranging from 4–11% in those 65–79 years old, we estimate a total of 480/2000 people will be screened from within this age group and between 19/480 and 53/480 people will be found to be in AF. In addition, it is estimated that the remaining 1520 customers present between 0.1–1.5% risk of AF suggesting between 2/1520 and 23/1520 additional presentations. In total, it is estimated that between 21 and 76 participants will be found to be suffering AF [27]. The prevalence of undiagnosed AF differs dependent on age. A systematic review of 30 single time-point AF screening studies including 122,571 participants (Lowres et al., 2013) reported the incidence of previously unknown AF as 1.0% (CI, 0.89–1.04%), increasing to 1.4% (CI, 1.2–1.6%) in those >65 years old. We recognise that we are recruiting a yonger population than those included in these studies, we therefore estimate that up to 28 people will be identified with undiagnosed AF.

2.1.4. Recruitment

A convenience sample of all those attending four designated supermarkets in the North West of England. We will station a researcher at the entrance of each store who will approach shoppers as they enter the store and invite them to participate in the study. Participant information sheets on the trolley will outline the study and provide details of how people can participate and also include details of the funder (Bristol Myers Squibb). The supermarkets included in the study are located within the Liverpool City Region, a region with pockets of high levels of deprivation. The supermarkets are situated within community hubs and recognise their social responsibility.

2.1.5. Informed Consent

Verbal consent will be gained prior to recruitment, with written consent obtained for all participants with an abnormal sensor reading whose personal data are required for onward referral to a consultant cardiologist.

2.1.6. Sensor Technology

The sensor used within this study, MyDiagnostick has been shown to be highly sensitive in detecting AF, with sensitivity levels ranging between 94% (95% CI 87–98%) [28] to 100% (95% CI 93–100%) [29]. The specificity of MyDiagnostick also compares well with other devices ranging between 93% (95% CI 85–97%) [28] to 95.9% (95% CI 91.3% to 98.1%) [29].

2.1.7. Study Procedures

We will recruit 2000 people attending one of four supermarkets in the North West of England. Each participant will use a supermarket trolley to undertake his or her routine shopping and, by doing so, will grip the trolley handle at differing points and for differing time periods. It will be made clear to each participant that only one person should push the trolley during his or her visit to the store. During each contact with the trolley handle, the sensor will assess the participant's pulse and store a recording of the rhythm strip, which is stored within the sensor's file storage system. These data will be downloaded alongside the personal data at the end of each day and deleted from the storage system. The handles of all trolleys will be sanitised between participants to minimise the risk of infection.

If the pulse sensor detects AF, the store pharmacist will be alerted by the researcher, and they will meet with the participant in the store to repeat the sensor check and undertake a manual pulse check. If the pulse sensor does not detect AF whilst the person is pushing the trolley, the researcher will meet with the participant and a manual pulse check will be undertaken once the person has completed their shopping. The procedure to be followed for each participant is outlined in Figure 4. An irregular pulse is defined as any irregularity between pulse waveforms in the radial artery within a period of sixty seconds. Guidance recommends pulse palpation as the first step for AF screening [10]. Two randomised controlled trials have found that pulse palpation is an effective and cost-effective approach for screening for AF [18,30]. All staff undertaking manual pulse checks will undertake additional training using simulation manikins to ensure that they are practising in line with the procedures outlined in the Royal Marsden Manual of Clinical Nursing Procedures [31]. However, a static sensor check (control) will provide an ability to differentiate between normal irregular pulses, e.g., sinus arrhythmia and more harmful conditions, including AF.

2.1.8. Outcome of Interest

The primary objective is to determine the effectiveness of a MyDiagnostick sensor embedded in the handle of a supermarket trolley in detecting AF.

The 12 lead ECGs of all participants referred for follow-up will be analysed for evidence of AF by a Consultant Cardiologist (GL). AF is characterized on the ECG by rapid, irregular waves that vary in size, form and timing. The results of this analysis will be compared with the participant's initial sensor recording and the static sensor recording to establish concordance.

Within this phase, there are two secondary objectives [1] to establish what percentage of the public will choose to carry out their shopping using a trolley with a pulse sensor in the handle and [2] to establish if shoppers, who have been identified as having an irregular pulse rate, would be willing to attend a local healthcare facility to have an ECG recorded.

The research team will introduce the new trolley to shoppers and the percentage uptake will be identified. These data will provide both the research team and the designers with valuable information that will inform the scalability of the device, if found to be effective.

With consent, details of all those who are found to have an irregular sensor reading will be stored on a research database. A Consultant Cardiologist will review the sensor recordings to eliminate those recordings that are uninterpretable and whereby the sensor has been activated by the presence of the artefact. These participants will be invited to return for an additional static check. The details of those with confirmed AF or suspected AF will be referred to a local healthcare facility. These details will be compared against those who attend a follow-up ECG. Anyone who has not attended an ECG assessment will be contacted by telephone within one week of referral. During this telephone call, the participant will again be made aware of the risks associated with an irregular pulse, and a further invitation to attend a clinic for an ECG will be offered. Participants who do not attend will be contacted on two occasions, after which, if they still have not attended, a letter will be sent to their GP explaining that the detection of an irregular pulse was found during a screening study. Follow-up care will then be provided by the GP as per routine NHS care. All data will be aggregated, enabling comparisons to be made. As the aim of the study is to assess the merit of the screening approach, the study follow-up will discontinue once the participant has attended the cardiology clinic or failed to attend despite two follow-up telephone calls. All treatment beyond this stage will be in line with contemporary NHS AF pathways, and consequently, the likely incidence of stroke will be extrapolated from the results of national datasets.

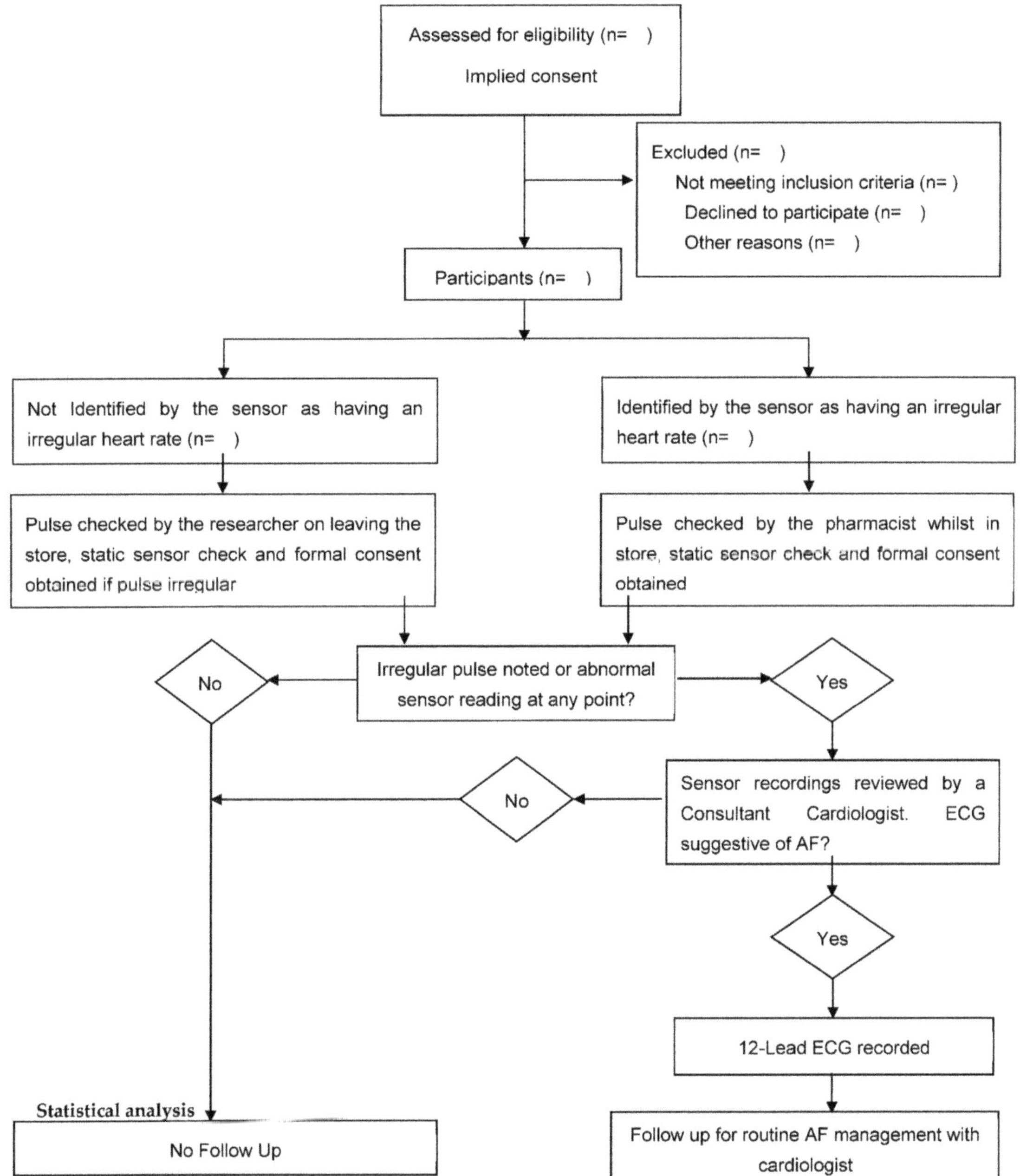

Figure 4. Phase 1 (Quantitative) procedure to follow regarding those participants to be recruited in a cross-sectional observational study determining the effectiveness of sensor technology to detect atrial fibrillation (AF) when embedded in the handles of supermarket trolleys. A minimum of two thousand participants will be recruited. A convenience sample will be used.

2.1.9. Statistical Analysis

A sensitivity of 95% and specificity of 90% is sufficiently accurate to be incorporated into clinical practice. If confirmed, they will make supermarkets with in-store pharmacies a nationwide relevant and cost-effective point of screening AF. Such high sensitivity and specificity values are also indicated as plausible by existing published papers, with 95% confidence intervals of 93–100% and 91.3–98.1% [29] and 87–98% and 85–97% [28], for sensitivity and specificity, respectively. In our proposed feasibility study, our precision calculations show that with between 21 and 76 AF cases, we will estimate the 95% confidence interval of sensitivity and specificity with interval widths of 19.0% and 9.8%. The most likely precision will be 12.4%.

Descriptive statistics including simple percentages will be estimated from this feasibility study to demonstrate uptake, rates of detection and successful referral. Generating these data will enable the researchers to determine the number of shoppers who need to be approached to recruit for an adequately powered future study. In addition, conditional percentages will be used to describe the sensitivity and specificity of both the sensor and the pulse check.

By recruiting from community supermarkets based in areas of high deprivation, it is likely that we will recruit a greater number of individuals who are known not to typically engage in traditional screening programmes. This approach attempts to correct the inverse care law, bringing healthcare to the community.

2.2. Phase 2

The second phase of the study involves an embedded qualitative approach. The primary objective is to explore the views of supermarket-based pharmacists and store managers on the merits of embedding pulse sensors in the handles of supermarket trolleys to enable the detection of AF. The secondary objective is to explore the pharmacists' perspectives and experiences of performing manual pulse checks on participants.

2.2.1. Informed Consent

The Senior Store Manager will identify all those staffs who are eligible for the study and will distribute a Participant Information Sheet and Consent Form accordingly. A member of the research team will be available to explain the study to potential participants at a mutually convenient time. All eligible participants will receive information about the study outlining the purpose, potential risks and implications of participation. They will be asked to sign and date an informed consent form that adheres to the ethical principles that have their origin in the Declaration of Helsinki prior to enrolment. The participant information sheet and informed consent form will acknowledge that the study data will be used in academic publications. However, participants will be informed that no personal or identifiable data will be shared outside the research team. All data will be anonymised prior to publication.

2.2.2. Study Design

Qualitative interviews will be undertaken to explore the views of pharmacists and store managers. All pharmacists and store managers from each supermarket will be invited to participate in a qualitative interview. Store managers are defined as any manager with responsibility for managing the store during the study period. Pharmacists are defined as being General Pharmaceutical Council (GPhC) registered and employed in the store. All participants must be able to provide written informed consent and devote time to attend the interview. The exclusion criteria include previous participation in the study in the event of a store manager or pharmacist transferring to a neighbouring store during the study period and absence from the store during the entirety of the data collection period.

2.2.3. Recruitment

A convenience sample including store managers and GPhC registered pharmacists employed within the supermarket within the study period will be recruited. Each store manager will identify potential participants.

2.2.4. Study Procedures

We will undertake individual semi-structured interviews with all store managers and pharmacists that have engaged in the screening study. The interviews will focus on exploring the supermarket employees' views on the merits of embedding sensors in the handles of the supermarket trolleys, the impact of using these trolleys to screen for AF on the workload of the pharmacists and the efficiency of the stores and identify any additional customer feedback that may have been received. The procedure for Phase 2 is outlined in Figure 5.

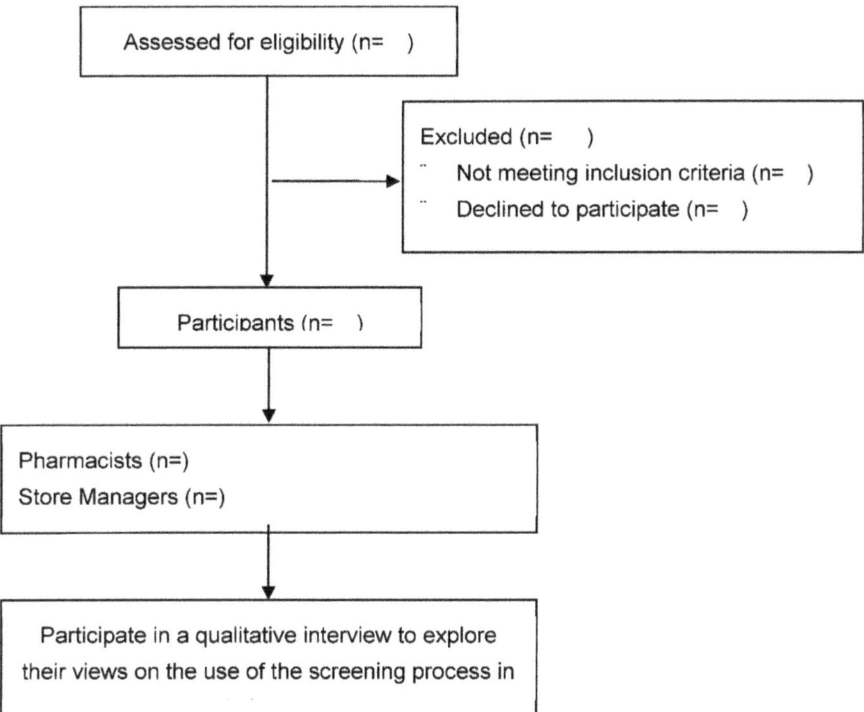

Figure 5. Phase 2 (Qualitative) procedure to follow when conducting semi-structured interviews exploring the views of supermarket-based pharmacists and store managers around the use of sensor technology to detect atrial fibrillation (AF) when embedded in the handles of supermarket trolleys.

2.2.5. Data Analysis

Thematic analysis using the qualitative data generated from the interviews with pharmacists and store managers will be independently analysed by two members of the research team adopting Braun and Clarke's (1996) framework for thematic analysis [32].

2.2.6. Patient and Public Involvement (PPI)

The Service Users for Research Endeavour (SURE) group at Liverpool Heart and Chest Hospital and the patient and public involvement group at the Liverpool Centre for Cardiovascular Science will review the research protocol, participant information sheet and consent form. Their advice will be sought throughout the design of the study.

2.2.7. Ethical Considerations

The ethical review has been granted by Liverpool John Moores University's University Research Ethics Committee. The study will be undertaken in compliance with the research protocol. During Phase 1, verbal consent will be obtained upon recruitment, with written consent secured for those with an abnormal sensor recording whose personal data will be required for onward referral for 12 lead ECGs. For the qualitative sub-study (Phase 2), written consent will be obtained for all participants.

3. Discussion

Many people with AF are undiagnosed, unaware and are subsequently at increased risk of stroke and mortality compared with people with symptomatic AF [33–35]. The reasons for these differences are unclear but are likely to occur because of limited preventative treatment, such as anti-coagulation and inadequate management of additional risk factors.

The high economic and individual burden of AF makes early identification through screening an attractive proposition to clinicians, patients and budget holders alike. However, trial data to support AF screening models are scarce [36,37]. Opportunistic screening has been found to be cost-effective [38] but remains reliant on a substantial health resource. Consequently, there is increasing interest in the use of patient-initiated ECG or photoplethysmography screening, with >100 mHealth apps and $\geq$400 wearable activity monitors reportedly available [39]. However, patient-initiated screening can be expensive to the consumer and consequently excludes those with lower household incomes. Moreover, this form of screening is dependent on individuals with few or no symptoms actively purchasing and participating in a screening ritual. Furthermore, the number of e-health applications is increasing exponentially with little or no scrutiny of their efficacy. It is therefore inevitable that the public will be exposed to poor-quality devices with limited accuracy, the results of which will lead to a combination of false reassurance for some and an inappropriate use of healthcare resources for others. Introducing sensor-based screening into everyday activities with immediate access to resident health care professionals could provide a means of capitalising on the improvements in technology whilst negating the need for individuals to invest in said technology and, in doing so, revolutionise the way that healthcare screening is offered.

This proposed study aims to introduce the concept of using sensor-based diagnostic technology in everyday activities whilst ensuring that all participants have access to a defined healthcare pathway, thus optimising the benefits that early diagnosis provides. The study design will allow the researchers to assess the acceptability of the concept, the efficacy of the sensors and the engagement of those found to be in AF. Finally, the study will address the uncertainties in optimal trial design to inform a future definitive trial.

Author Contributions: I.D.J.: Conceptualisation, methodology, validation, original draft preparation, writing, reviewing and editing, supervision and funding acquisition; D.A.L.: conceptualisation, methodology, writing, reviewing and editing, and funding acquisition; R.R.L.: methodology, investigation, formal analysis, writing, reviewing and editing; D.O.: writing, reviewing and editing; L.N.: writing, reviewing and editing; P.E.P.: conceptualisation, methodology, writing, reviewing and editing; G.C.: conceptualisation, software, formal analysis, data curation, writing, reviewing and editing; A.S. (Andy Shaw): methodology, software, validation, resources; E.J.S.: resources, data curation, supervision, project administration, writing, reviewing and editing; A.S. (Aimeris Santos): formal analysis, investigation, writing, reviewing and editing, visualisation; E.E.M.: formal analysis, investigation, writing, reviewing and editing, visualisation; A.A.: formal analysis, investigation, writing, reviewing and editing, visualisation; N.T.: formal analysis, investigation, writing, reviewing and editing, visualisation; G.Y.H.L.: conceptualisation, methodology, software, validation, formal analysis, writing, reviewing and editing, project administration, funding acquisition. All authors have read and agreed to the published version of the manuscript.

Funding: The authors would like to thank Bristol Myers Squibb for their financial support (funding number REQ-0000022262), which has enabled this novel study to be undertaken.

Informed Consent Statement: Informed consent will be obtained from all subjects involved in the study.

Conflicts of Interest: D.A.L. has received investigator-initiated educational grants from Bristol-Myers Squibb (BMS), has been a speaker for Bayer, Boehringer Ingelheim and BMS/Pfizer and has consulted for BMS and Boehringer Ingelheim. P.E.P. owns shares in AstraZeneca PLC and has received honoraria and/or travel reimbursement for events sponsored by AKCEA, Amgen, AMRYT, Link Medical, Mylan, Napp, Sanofi. L.N. has received speaker fees from Pfizer, BMS and Daiichi Sankyo. G.Y.H.L. is a consultant and speaker for BMS/Pfizer, Boehringer Ingelheim and Daiichi-Sankyo. No fees are received personally. There are no other conflicts of interest.

References

1. British Heart Foundation. Cardiovascular Disease Statistics. 2018. Available online: https://www.bhf.org.uk/what-we-do/our-research/heart-statistics/heart-statistics-publications/cardiovascular-disease-statistics-2018 (accessed on 6 February 2019).
2. Xu, X.-M.; Vestesson, E.; Paley, L.; Desikan, A.; Wonderling, D.; Hoffman, A.; Da Wolfe, C.; Rudd, A.G.; Bray, B.D. The economic burden of stroke care in England, Wales and Northern Ireland: Using a national stroke register to estimate and report patient-level health economic outcomes in stroke. *Eur. Stroke J.* **2018**, *3*, 82–91. [CrossRef] [PubMed]
3. Newton, J.N.; Briggs, A.D.; Murray, C.J.L.; Dicker, D.; Foreman, K.J.; Wang, H.; Naghavi, M.; Forouzanfar, M.H.; Ohno, S.L.; Barber, R.M.; et al. Changes in health in England, with analysis by English regions and areas of deprivation, 1990–2013: A systematic analysis for the Global Burden of Disease Study 2013. *Lancet* **2015**, *386*, 2257–2274. [CrossRef]
4. Patel, A.; Berdunov, V.; Quayyum, Z.; King, D.; Knapp, M.; Wittenberg, R. Estimated societal costs of stroke in the UK based on a discrete event simulation. *Age Ageing* **2020**, *49*, 270–276. [CrossRef] [PubMed]
5. Hannon, N.; Sheehan, O.; Kelly, L.; Marnane, M.; Merwick, A.; Moore, A.; Kyne, L.; Duggan, J.; Moroney, J.; McCormack, P.M.; et al. Stroke associated with atrial fibrillation—Incidence and early outcomes in the north Dublin population stroke study. *Cerebrovasc. Dis.* **2010**, *29*, 43–49. [CrossRef] [PubMed]
6. Marini, C.; De Santis, F.; Sacco, S.; Russo, T.; Olivieri, L.; Totaro, R.; Carolei, A. Contribution of atrial fibrillation to incidence and outcome of ischemic stroke: Results from a population-based study. *Stroke* **2005**, *36*, 1115–1119. [CrossRef] [PubMed]
7. Burdett, P.; Lip, G.Y.H. Atrial Fibrillation in the United Kingdom: Predicting Costs of an Emerging Epidemic Recognising and Forecasting the Cost Drivers of Atrial Fibrillation-related costs. *Eur. Heart J. Quul. Care Clin. Outcomes* **2020**, *8*, 187–194. [CrossRef]
8. NHS digital Quality and Outcomes Framework 2020/2021. Available online: https://fingertips.phe.org.uk/search/atrial%20fibrillation (accessed on 1 February 2022).
9. Lip, G.Y.; Nieuwlaat, R.; Pisters, R.; Lane, D.A.; Crijns, H.J. Refining clinical risk stratification for predicting stroke and thromboembolism in atrial fibrillation using a novel risk factor-based approach: The euro heart survey on atrial fibrillation. *Chest* **2010**, *137*, 263–272. [CrossRef]
10. Hindricks, G.; Potpara, T.; Dagres, N.; Arbelo, E.; Bax, J.J.; Blomström-Lundqvist, C.; Boriani, G.; Castella, M.; Dan, G.-A.; Dilaveris, P.E.; et al. 2020 ESC Guidelines for the diagnosis and management of atrial fibrillation developed in collaboration with the European Association for Cardio-Thoracic Surgery (EACTS): The Task Force for the diagnosis and management of atrial fibrillation of the European Society of Cardiology (ESC) Developed with the special contribution of the European Heart Rhythm Association (EHRA) of the ESC. *Eur. Heart J.* **2021**, *42*, 373–498. [CrossRef]
11. Hart, R.G.; Pearce, L.A.; Aguilar, M.I. Meta-analysis: Antithrombotic therapy to prevent stroke in patients who have nonvalvular atrial fibrillation. *Ann. Intern. Med.* **2007**, *146*, 857–867. [CrossRef]
12. Ruff, C.T.; Giugliano, R.P.; Braunwald, E.; Hoffman, E.B.; Deenadayalu, N.; Ezekowitz, M.D.; Camm, A.J.; Weitz, J.I.; Lewis, B.S.; Parkhomenko, A.; et al. Comparison of the efficacy and safety of new oral anticoagulants with warfarin in patients with atrial fibrillation: A meta-analysis of randomised trials. *Lancet* **2014**, *383*, 955–962. [CrossRef]
13. Potpara, T.S.; Lip, G.Y.H.; Blomstrom-Lundqvist, C.; Boriani, G.; Van Gelder, I.C.; Heidbuchel, H.; Hindricks, G.; Camm, A.J. The 4S-AF Scheme (Stroke Risk; Symptoms; Severity of Burden; Substrate): A Novel Approach to In-Depth Characterization (Rather than Classification) of Atrial Fibrillation. *Thromb. Haemost.* **2021**, *121*, 270–278. [CrossRef] [PubMed]
14. Romiti, G.F.; Pastori, D.; Rivera-Caravaca, J.M.; Ding, W.Y.; Gue, Y.X.; Menichelli, D.; Gumprecht, J.; Kozieł, M.; Yang, P.-S.; Guo, Y.; et al. Adherence to the 'Atrial Fibrillation Better Care' Pathway in Patients with Atrial Fibrillation: Impact on Clinical Outcomes—A Systematic Review and Meta-Analysis of 285,000 Patients. *Thromb. Haemost.* **2021**, *122*(3), 406–414. [CrossRef] [PubMed]
15. Fitzmaurice, D.A.; Hobbs, F.D.R.; Jowett, S.; Mant, J.; Murray, E.T.; Holder, R., Raftery, J.P.; Bryan, S.; Davies, M.; Lip, G.Y.H.; et al. Screening versus routine practice in detection of atrial fibrillation in patients aged 65 or over: Cluster randomised controlled trial. *BMJ* **2007**, *335*, 383. [CrossRef] [PubMed]
16. Dilaveris, P.E.; Kennedy, H.L. Silent atrial fibrillation: Epidemiology, diagnosis, and clinical impact. *Clin. Cardiol.* **2017**, *40*, 413–418. [CrossRef]
17. Freedman, B.; Camm, J.; Calkins, H.; Healey, J.S.; Rosenqvist, M.; Wang, J.; Albert, C.; Anderson, C.S.; Antoniou, S.; Benjamin, E.J.; et al. Screening for Atrial Fibrillation: A Report of the AF-SCREEN International Collaboration. *Circulation* **2017**, *135*, 1851–1867. [CrossRef]
18. Hobbs, F.D.; Fitzmaurice, D.A.; Mant, J.; Murray, E.; Jowett, S.; Bryan, S.; Raftery, J.; Davies, M.; Lip, G. A randomised controlled trial and cost-effectiveness study of systematic screening (targeted and total population screening) versus routine practice for the detection of atrial fibrillation in people aged 65 and over. The SAFE study. *Health Technol. Assess.* **2005**, *9*, iii–iv, ix–x, 1–74. [CrossRef]

19. Department of Health and Social Care. Pharmacy in England: Building on Strengths—Delivering the Future. 2008. Available online: https://www.gov.uk/government/publications/pharmacy-in-england-building-on-strengths-delivering-the-future (accessed on 6 February 2019).
20. Lowres, N.; Neubeck, L.; Salkeld, G.; Krass, I.; McLachlan, A.J.; Redfern, J.; Bennett, A.A.; Briffa, T.; Bauman, A.; Martinez, C.; et al. Feasibility and cost-effectiveness of stroke prevention through community screening for atrial fibrillation using iPhone ECG in pharmacies. *The SEARCH-AF study. Thromb. Haemost.* **2014**, *111*, 1167–1176. [CrossRef]
21. Willis, A.; Rivers, P.; Gray, L.J.; Davies, M.; Khunti, K. The effectiveness of screening for diabetes and cardiovascular disease risk factors in a community pharmacy setting. *PLoS ONE* **2014**, *9*, e91157. [CrossRef]
22. Lowres, N.; Krass, I.; Neubeck, L.; Redfern, J.; McLachlan, A.J.; Bennett, A.A.; Freedman, B. Atrial fibrillation screening in pharmacies using an iPhone ECG: A qualitative review of implementation. *Int. J. Clin. Pharm.* **2015**, *37*, 1111–1120. [CrossRef]
23. Statista. Average Number of Morrisons Customers per Store per Week in the United Kingdom from Financial Year 2009/2019 to 2020/2021. 2021. Available online: https://www.statista.com/statistics/382353/morrisons-weekly-customer-numbers-united-kingdom-uk/ (accessed on 4 December 2019).
24. Public Health England. Atrial Fibrillation Prevalence Estimates for Local Populations. 2015. Available online: https://www.gov.uk/government/publications/atrial-fibrillation-prevalence-estimates-for-local-populations (accessed on 4 December 2019).
25. Public Health England. Public Health Profiles. 2021. Available online: https://fingertips.phe.org.uk/search/atrial#page/0/gid/1/pat/46/par/E39000026/ati/154/are/E38000056 (accessed on 4 December 2019).
26. Carpenter, J.M.; Moore, M. Consumer demographics, store attributes, and retail format choice in the US grocery market. *Int. J. Retail Distrib. Manag.* **2006**, *34*, 434–452. [CrossRef]
27. Lowers, N.; Neubeck, L.; Redfern, J.; Freedman, S. Screening to identify unknown atrial fibrillation. *Thromb Haemost.* **2013**, *110*, 213–222. [CrossRef] [PubMed]
28. Vaes, B.; Stalpaert, S.; Tavernier, K.; Thaels, B.; Lapeire, D.; Mullens, W.; Degryse, J. The diagnostic accuracy of the MyDiagnostick to detect atrial fibrillation in primary care. *BMC Fam. Pract.* **2014**, *15*, 113. [CrossRef] [PubMed]
29. Tieleman, R.G.; Plantinga, Y.; Rinkes, D.; Bartels, G.L.; Posma, J.L.; Cator, R.; Hofman, C.; Houben, R. Validation and clinical use of a novel diagnostic device for screening of atrial fibrillation. *Europace* **2014**, *16*, 1291–1295. [CrossRef] [PubMed]
30. Morgan, S.; Mant, D. Randomised trial of two approaches to screening for atrial fibrillation in UK general practice. *Br. J. Gen. Pract.* **2002**, *52*, 373–374, 377–380. Available online: https://pubmed.ncbi.nlm.nih.gov/12014534/ (accessed on 6 February 2019).
31. Dougherty, L. *The Royal Marsden Manual of Clinical Nursing Procedures*, 9th ed.; Lister, S., West-Oram, A., Eds.; Wiley-Blackwell: Chichester, UK, 2015.
32. Braun, V.; Clarke, V. Using thematic analysis in psychology. *Qual. Res. Psychol.* **2006**, *3*, 77–101. [CrossRef]
33. Boriani, G.; Laroche, C.; Diemberger, I.; Fantecchi, E.; Popescu, M.I.; Rasmussen, L.H.; Sinagra, G.; Petrescu, L.; Tavazzi, L.; Maggioni, A.P.; et al. Asymptomatic atrial fibrillation: Clinical correlates, management, and outcomes in the EORP-AF Pilot General Registry. *Am. J. Med.* **2015**, *128*, 509–518.e2. [CrossRef]
34. Potpara, T.S.; Polovina, M.M.; Marinkovic, J.M.; Lip, G.Y. A comparison of clinical characteristics and long-term prognosis in asymptomatic and symptomatic patients with first-diagnosed atrial fibrillation: The Belgrade Atrial Fibrillation Study. *Int. J. Cardiol.* **2013**, *168*, 4744–4749. [CrossRef]
35. Siontis, K.C.; Gersh, B.J.; Killian, J.M.; Noseworthy, P.A.; McCabe, P.; Weston, S.A.; Roger, V.L.; Chamberlain, A.M. Typical, atypical, and asymptomatic presentations of new-onset atrial fibrillation in the community: Characteristics and prognostic implications. *Heart Rhythm* **2016**, *13*, 1418–1424. [CrossRef]
36. Steinhubl, S.R.; Waalen, J.; Edwards, A.M.; Ariniello, L.M.; Mehta, R.R.; Ebner, G.S.; Carter, C.; Baca-Motes, K.; Felicione, E.; Sarich, T.; et al. Effect of a Home-Based Wearable Continuous ECG Monitoring Patch on Detection of Undiagnosed Atrial Fibrillation: The mSToPS Randomized Clinical Trial. *JAMA* **2018**, *320*, 146–155. [CrossRef]
37. Welton, N.J.; McAleenan, A.; Thom, H.H.; Davies, P.; Hollingworth, W.; Higgins, J.P.; Okoli, G.; Sterne, J.; Feder, G.; Eaton, D.; et al. Screening strategies for atrial fibrillation: A systematic review and cost-effectiveness analysis. *Health Technol. Assess.* **2017**, *21*, 1–236. [CrossRef]
38. Nielsen, J.C.; Lin, Y.J.; de Oliveira Figueiredo, M.J.; Sepehri Shamloo, A.; Alfie, A.; Boveda, S.; Dagres, N.; Di Toro, D.; Eckhardt, L.L.; Ellenbogen, K.; et al. European Heart Rhythm Association (EHRA)/Heart Rhythm Society (HRS)/Asia Pacific Heart Rhythm Society (APHRS)/Latin American Heart Rhythm Society (LAHRS) expert consensus on risk assessment in cardiac arrhythmias: Use the right tool for the right outcome, in the right population. *J. Arrhythmia* **2020**, *36*, 553–607. [CrossRef]
39. Li, K.H.C.; White, F.A.; Tipoe, T.; Liu, T.; Wong, M.C.; Jesuthasan, A.; Baranchuk, A.; Tse, G.; Yan, B.P. The Current State of Mobile Phone Apps for Monitoring Heart Rate, Heart Rate Variability, and Atrial Fibrillation: Narrative Review. *JMIR Mhealth Uhealth* **2019**, *7*, e11606. [CrossRef] [PubMed]

MDPI
St. Alban-Anlage 66
4052 Basel
Switzerland
Tel. +41 61 683 77 34
Fax +41 61 302 89 18
www.mdpi.com

Journal of Personalized Medicine Editorial Office
E-mail: jpm@mdpi.com
www.mdpi.com/journal/jpm

www.ingramcontent.com/pod-product-compliance
Lightning Source LLC
LaVergne TN
LVHW070653170726
843515LV00004B/401